GETTING HERE

The Story of Human Evolution

• New Edition •

Books by William Howells

Mankind So Far
The Heathens
Back of History
Mankind in the Making
The Pacific Islanders
Evolution of the Genus Homo
Getting Here

GETTING HERE

The Story of Human Evolution

• New Edition •

by

William Howells

THE COMPASS PRESS
WASHINGTON, DC

Illustrations by Ann Meagher-Cook
Cover design by Janet West Garrett
Book design and typesetting by Gary Roush
Printed and bound in the U.S. by Edwards Brothers

Cover photo of Cro Magnon skull from Origines de l'Homme, *courtesy of the Laboratoire de Préhistoire of the Musée de l'Homme, Paris.*

Library of Congress Cataloguing-in-Publication Data

Howells, W.W. (William White), 1908–
    Getting here: the story of human evolution — by William Howells.
    New ed.
        p. cm.
    Includes index.
    ISBN 0-929590-16-3 (cloth: alk. paper)
    ISBN 0-929590-17-1 (paper: alk. paper)
    1. Human evolution. I. Title
GN281.H685 1997
599.93'8—dc21

                                                              97-7443
                                                                CIP

The Compass Press is an imprint of Howells House
Box 9546 Washington, DC 20016

To all the friends, living and dead,
on whose work and thought I have relied,
this book is admiringly dedicated.

Table of Contents

List of Illustrations

PHOTOS

Preface

Half a century ago, writing a book about human evolution was simpler. We had the bones of a corporal's guard of "early men," we knew the anatomy of the apes and the monkeys and other primates, and we knew about Darwin and something about genes. By and large, that was it. The writer did not have to keep a reader treading water while he explained complicated matters and gave balanced viewpoints.

I have before me some pages of the venerable *Illustrated London News* from 1929. They show an artist's as-in-life recreations of the fossil men, made under the eye of Professor Elliot Smith, a world authority. One is the Java Man, and a second is the Peking Man, whose first skull, found two months earlier, was the occasion for the article. The last reconstruction is the Piltdown Man, who never existed; in 1929 it was not known that his remains were bogus.

Seen today, these reconstructions seem pretty bad. They are really only icons. They simply declared that, at some time in the past, members of the human family were more primitive and smaller-brained than we are. Ages of these few ancients were not known, and family trees based on them were pure speculation, revealing nothing. The fossils were like posters on the wall, announcing that the circus was coming. Today, the circus is in town.

Here is what we now see of the past. Our particular kind of animal came on the scene in the last one-thousandth part of earth's history, four to five million years ago. At the moment, we perceive four general stages, each of the first three lasting something over a million years before giving rise to the next. These stages are:

- *Australopithecus*, who had, roughly speaking, our teeth and feet;
- *Homo habilis*, who made simple stone tools and had the first slightly larger brain;
- *Homo erectus*, a brain still larger and a skeleton much like our own;
- *Homo sapiens*, all today's people, with large brains and complex culture.

Fig. 1. Dubious Ancestors
These 1929 attempts to restore "early men" ranged from 75 percent to 100 percent imaginary. No fossil faces had been found for the Java and Peking Men (top and middle); Piltdown Man (bottom) was fraudulent.

Thus, there is now a structure in place, not just icons. We have a large and growing army of fossil forerunners and, equally important, we know roughly when each of them lived. We know more and more about what the world was like at different times. And there are ingenious and persuasive new methods of close study and of broad comparison of what we have from the ground. The scholars argue heatedly about this connection or that, but they know what they are arguing about, factoring in all the background of time, anatomy, geography, past climates, and evolutionary processes.

From time to time someone says that, because they are dealing with our own species, the anthropologists are too closely and emotionally involved to get it all straight. That is hardly fair; it would disqualify them altogether and, in any case, no kind of scientist really has icewater in his veins instead of blood. But it should be possible both to review our evolution, the most interesting of histories, while also standing a little to one side watching the anthropologists at work.

That is why I have written this book. I would like to tell an interested reader not only the known story but also why the anthropologists think what they think and argue as they argue. Now is a good time to view the web of facts, ideas, and reasons for controversy, as the guide to looking at future discoveries. Of course, no time is the perfect time because discoveries are constant, and no book can be a final book. It is less than five years since the first edition of this one appeared, and important facts, of dating as much as fossils, have called for considerable changes of later parts. This is more than mere additions and corrections. The new information has led to a current of new opinion among various authors, and therefore much of what appears herein represents changed formulations of view — certainly in my own case — in important ways. All this goes to remind us how refinements of understanding come along constantly. And it shows how a new find may carry an ever bigger load of satisfaction and information as it takes its place in the whole landscape of our knowledge. Certainly the highways in this landscape have been found — that is the significant change from the old days.

Chapter 1

The Evolution of Evolution

When today becomes yesterday, people will look back on our time as one when the sciences of nature began to move forward at a new pace. Like passengers in a plane, we ourselves may not sense the motion. But it is momentous.

Early in the present century, scientific progress had been picking up speed. Yet it is only since World War II that astronomers have seen the great walls of uncountable galaxies or the violent explosions of distant stars taking place in that calm black night. Only since then have geologists found the reasons for earthquakes and volcanoes by tracking the movements of continental plates. And until then biologists were still ignorant of the structure of DNA, the stuff of genes and chromosomes. In fact, they were wrong in the middle 1950s — just yesterday — about the number of chromosomes in the cells of our own well-studied species. Today the molecular sequences of genes and chromosomes can be read in detail, and genetic engineering is becoming almost commonplace.

Knowledge of our evolutionary past has also gone ahead, if not in so breath-taking a way. The sciences named above have been powered by equally imposing progress in technology: astronomers have space telescopes, and geologists have satellites and ocean-floor maps. But paleontologists have no such commanding ways of making fossils of the right kind and time appear at their bidding; as always, they must patiently keep looking in likely places. Paleontology is an arduous outdoor science.

Nevertheless, the numbers and kinds of fossils have been multiplying greatly. And, very important, the findings in other branches of biology — specifically the molecular kind involving DNA — have made themselves felt. Because of these new things, we are changing the ways we think about human evolution. And we must take account of more complex evolutionary processes as we know them from zoology and genetics, while we also try to see past ancestors in their actual climates and to understand what kinds of behavior led to changes — standing up and walking, for example. In other words, we now have a vast background of biology into which our own existence has to be fitted.

There is another kind of change. Nowadays we deal with fossils not simply as though they were famous people — "Java Man," "Peking Man" — but rather by considering situations of the past in their entirety. We have long known about Cro Magnon Man as a "caveman," but the intense discussion today is about a far broader problem: when and how the Cro Magnons, people of our own kind, appeared in the first place. Also, at a much earlier time, about ten million years ago, we are seeing an expanding menagerie of extinct apes, newly discovered; and we are asking how human beings emerged from among them. These are problems that engage large numbers of today's anthropologists.

Theories and Hypotheses

It is commonly said, by way of easy dismissal, that evolution is "just a theory." This springs from a lack of clarity about scientific thinking and scientific progress. "Theory" really means the whole developed network of ideas in a given science: the well-established facts, the probabilities, and the possibilities to which these lead. Atomic theory may be the best example. But people are apt to say "theory" when they mean "hypothesis." A hypothesis is one of those probabilities or possibilities on the edge of theory, which may or may not explain a set of facts or observations. Usually two, or several, hypotheses compete. One or more is shown by further work to be wrong, but a hypothesis that stands the tests will become part of theory.

Picture all scientific theory as a single building. It has different wings — astronomical theory, atomic theory, biological theory, and others — but they join together at the center, where the building blocks fit solidly. At some points you can go from one wing to another. The building constantly grows around the outside, and the structure is loose in many places, where new pieces (hypotheses) may fit only tentatively. You are at liberty to carve a new hypothesis to suit yourself, but if it holds that the moon is made of green cheese, there is no possible place for it to fit. If the hypothesis has to do with extra-terrestrial space ships, it may captivate large numbers of people, but attaching it to the main structure will be a problem.

On the other hand, some new pieces will fit very well. The hypothesis of drifting continents did not find a place when it was first carved, but then the geological wing of the building gradually changed and grew in such a way that the hypothesis fitted perfectly and attracted new blocks. And Darwin crafted a whole segment of several important blocks that soon became the central part of the biological wing of theory, to which all new blocks have had to be accommodated ever since.[1]

No astronomer has yet seen black holes (by theory they can't be seen, anyhow). But the hypothesis of their existence is very strong. Astronomers and physicists can't escape that conclusion because of so much else they increasingly see in galaxies, and because black holes are quite at home in the theory of relativity. So the *hypothesis* of black holes is becoming a part of the *theory*.

Evolutionary theory is another such body of ideas and understanding. It does not lend itself to experiment and instrumentation like atomic theory, which supercolliders are built to test. And nobody, they say, has seen anything evolve. Not true: for example, the main malaria parasite has, in the last thirty years, changed its gene structure to become immune to the best anti-malaria drugs. That is classic textbook evolution. The simple fact is that, today, not only fossils and everything we know about kinds of animals and plants, but also all our work with genes or the testing of viruses on unfortunate chimpanzees — none of these things make any sense

---

1. It may be helpful to think of religious understanding of the universe and life, not as a building that may have loose pieces under your feet but as a great painting. It is beautiful, it is unified, it gives explanation and meaning, and it is soul-satisfying to viewers and important to society at large. It was painted to last, and it does not change, and that is its difference from science. There is no reason why painting and building should not co-exist, although they do not need each other.

After writing this I noted a letter to *Nature* which says it better, from Professor Cesare Emiliani, to whom we owe chronological ordering of marine sediments by oxygen isotopes, a major achievement. In the letter he deplored the use of radiocarbon in dating the Shroud of Turin, bearing the imprint of the dead Christ, to the fourteenth century A.D. He wrote in part: "Religion is perfect and unchangeable, the work of God. Science is imperfect, and, I suspect, the work of the Devil. The two should never be mixed. The scientists who participated in the dating of the Shroud of Turin should repent and promise never to do anything like that again. Creationists are even more guilty, for they have been mixing science and religion for years and years. They should abandon their evil practices forthwith, lest the wrath of God descend upon them like a ton of bricks."

without evolution. Evolutionary theory is now the center of the whole science of biology.

This does not mean that everything about evolutionary history will soon be neatly wrapped up. A very great deal is known as fact, but a very great deal more still is not, and the scientific journals are full of contention. This book will deal as much with the latter as with the former. But, little by little, this or that controversy becomes resolved, and we can turn to new ones.

Here is an example of a good hypothesis. In 1871, after noting that human beings shared the essential traits of the Old World monkeys and apes, Charles Darwin wrote:

> It is therefore probable that Africa was formerly inhabited by extinct apes closely allied to the gorilla and chimpanzee; and as these two species are now man's nearest allies, it is somewhat more probable that our early progenitors lived on the African continent than elsewhere.

That is typical of the cautious Darwin. He was not guessing. For his own day he knew a lot about the world's animals and about human and primate anatomy, but there were almost no fossil ancestors on the shelves. Twenty years later, as it happened, the first really primitive human fossil was found in Java, and important ape fossils then came to light in Pakistan. So less cautious writers, with an Asian mind-set into the bargain, began arguing for Asia as our first home. Generous funds were spent to prove it. But Darwin was right, as has been resoundingly shown by all our recent studies of ape biochemistry and by the plentiful human fossils from East and South Africa.

There is a march in understanding nature, which has become a rush in recent years. In the early seventeenth century, Galileo was popularizing the hypothesis that the earth went round the sun, not the opposite. The Inquisition did not respond that this was "just a theory" — they told him to take it back, or else. He did (to protect his friends).[2] In the next century, the great French naturalist Comte de Buffon (a member of the Academy of Sciences at the age of twenty-six) wrote a many-volume treatise on natural science. In the

---

2. One influential friend tried to get a revocation of Galileo's sentence of house arrest. The Pope told the friend to his face: "The Holy Office is not in the habit of hearing defenses." In 1979 Pope John Paul II acknowledged that the Church had been in error.

early part, his views on the history of the earth attracted the notice of the Faculty of Theology, which found them reprehensible and contrary to the principles of religion. They told Buffon he should retract. He did.

In a later volume he wrote that, considering their close anatomical likeness, one might suppose that by a process of variation and degeneration, donkeys as a species had arisen from horses, and likewise apes from men; and given this, one might imagine that all animals had come, by the same kind of change, from a single species. That idea, one species rising from another, is evolution, plain and simple. But Buffon did not wait to hear from the theologians this time. He went on: "Mais non; it is clear from Scripture that all animals participated equally in the grace of creation, and that the first two of every species came fully formed from the hands of the Creator."

Darwin's Gift

A century later, in liberal Victorian England, Charles Darwin faced no such dictates from religious authority. What he faced, along with some immediate support, was simply indignant opposition from clergy, from much of the public, and from many of his fellow naturalists. These last were typically religious men and, given their own views of animal species, were more comfortable with Biblical creation than with godless natural forces. Each living species, in the accepted view, had been created as it now appears: none changed with time, none became extinct, and no new ones arose. Yes, of course, there were varieties within a species, and individuals varied from one to another. Still, the typical form of any species was what God had in mind.

But Darwin perceived that the variation was crucial. His achievement was colossal: he put evolutionary theory on its feet and gave a main explanation of its workings. He quickly persuaded most of his colleagues, and yet it was decades before they understood in detail how right he was.

With his phenomenal powers of observation, his curiosity, and his wide experience with fossils and living forms, he saw much more deeply and thoughtfully than his peers. In South America, he perceived strong evidence of change and of cases of extinction and, particularly in the Galapagos Islands, he noted sets of closely related species whose nearest relatives were on the nearest continent, South

America. He came at last to the conviction that species did arise and change in a pattern suggesting branchings from a parent form, a pattern that could not be explained by creation, as he himself had once thought. Indeed, he said that all life could have come from a single original species — Buffon's passing fancy became the central fact.

In the same way, Darwin was more impressed with natural variation and was thoroughly acquainted with varieties in domestic animals and plants; and he began his key work, *On the Origin of Species,* with a discussion of variation under domestication. In fact, though it was right under everyone's nose, nobody had seen that breeders were actually performing small-scale evolution, although not to the point of producing separate new species. In the single species of dogs, for example, the varied breeds were produced by careful and constant selection of those traits of shape, color, and behavior that the breeders wanted.

To a number of minds, evolution in the simple sense of change had already begun to seem possible. The great French naturalist Lamarck had made the case early in the nineteenth century. But what made evolution work? An answer was waiting to be given, and Darwin gave it: natural selection. It was stunningly simple. Nature does what animal breeders do, but gradually, and over longer periods of time. In a species, those individuals who are best adapted in their hereditary endowment will succeed, and survive, better than their fellows. That is the "struggle for existence." Those successful individuals will leave more offspring, and this will move the mode, the central tendency of the species, in their direction. Following generations will continue to undergo selection for still better adaptation: running, swimming, seeing, hiding, chewing, or all together. So the species is fluid, ready to change and, if two different countrysides are available, to divide so as to occupy both successfully.

That is adaptation by selection, Darwinism in a nutshell. Understanding variation, Darwin understood that the more "fit"[3]

---

3. It is important to remember that "fitter" does not mean "better" or "more advanced," but only "more likely to survive." Take an often-cited demonstration of evolution at work. In many parts of England there are populations of moths of a light speckled coloration, populations which also contain a less common black genetic variant. Following the Industrial Revolution, whose pollution blackened tree trunks, the black moths became the more common form, replacing the

were born already more fit. In this he differed from his predecessor Lamarck, who thought that they became more fit during life and somehow passed on such acquired fitness. But Darwin was greatly vexed by one problem: how did the heritable variation, the raw material, arise to begin with? In particular, he was perplexed as to how enhanced fitness or adaptation was maintained. In natural conditions, why should it be that a slightly better adaptation, occurring in one or a few individuals, was not simply lost in a generation or so, through blending with more ordinary members of the population? The key had, in fact, been found by Gregor Mendel in Bohemia, but if Darwin ever read Mendel's paper, it did not sink in. Darwin was not the only one who failed to notice Mendel's work.

Mendel discovered what we now call genes. Genes are incompressible. In the simplest case, if a gene is "recessive" (Mendel's prime finding), meaning that its effect is suppressed by its "dominant" partner gene, then the effect will not appear in that generation. But the effect is not diminished or lost: the gene itself survives. Genes do not "blend": they combine. You are born with genes, and especially with new combinations of genes, and this is what brings about variation and the possibility of advantageous variations which are also hereditary — just what Darwin wanted. So selection favors the fitter individual, but essentially it is selecting the fitter's genes, which thus become more numerous in the population.

Working on garden peas and taking single pairs of genes separately, Mendel was playing the piano with one finger, not realizing he had invented the whole instrument. Each key makes its own unvarying plink and does not show how music of infinite variety can be made by organizing the whole. That is what we have arrived at now. Mendel's findings lay dormant from 1866 to 1900, when they were rediscovered. In this century they have become the giant science of genetics.

---

speckled; in other words, the population had become typically black. A controlled experiment at one place near Liverpool showed that birds ate the moths they could see best: black moths on natural bark, speckled on blackened bark. With pollution being recently reduced, the trend has been reversing. This is pure sideways evolution. It is also a demonstration of the Red Queen principle. In *Through the Looking Glass,* Alice found the Red Queen running as hard as she could just to stay in the same place. In the real world, an environment may change — temperature, humidity, parasites, food, and so on, in this case, concealment — obliging a species to adapt or perish.

One more essential element was early recognized: mutations. Genes normally remain exactly the same in reproducing themselves but may, rarely, mutate or accidentally change in constitution and thus in effect. Mutations are a main source of new heritable material, even though much of it may be harmful.

The Rest of the Apple: Patterns and Processes

Darwin and Mendel are the core, the essentials of understanding. These basics work together. The gene pool — the hereditary property of a population of animals — maintains the variation of the population or species, and mutation tends to increase that variation. Darwin's selection cuts back the less favorable variation, in that way sculpturing the inheritance of the species.

Naturally, these are the simplest terms. Today, there is vastly more to evolutionary theory, most of which we can take in stride. Particularly important are matters of pattern, seen primarily in fossils. There are two main kinds of pattern, not necessarily exclusive.

First, gradual change. A species evolves, sometimes extremely slowly, sometimes rapidly if it must adapt to a changing environment or perish. Over time, an older species has changed into a new species. Or so paleontologists say, looking at the bones and seeing a difference as great as that between two living species. And so they award separate species names to the earlier and the later.

However, there are twin difficulties. Two living species can be told apart because they differ enough in the structuring of their genes to inhibit or prevent their interbreeding. But fossils cannot be bred. Also, the transition is apt to be gradual: if you had the whole evolutionary series, it would be difficult to point to a line between the old species and the new. Actually, we almost never have something like such a continuous series, and the problem of separation seldom arises. But wait: in our own case it does, at various points.

Gradual change is essentially what Darwin had in mind, and in fact is often called Darwinian evolution for short.[4] The second

---

4. So completely smooth a kind of change does not actually seem to be typical; instead, a pattern of "punctuated equilibria" is suggested by some paleontologists. In this, evolution proceeds by fits and starts, for reasons not clear. This can be seen when good time series of numerous fossils, like snails, can be studied. Long periods of stasis are interrupted by episodes of rapid change, when the species is

kind of pattern is the splitting of species into segments, giving rise not to a stately family tree but to an unruly bush. If a species is divided into isolated segments in differing environments, these smaller segments can evolve more rapidly than a large population that is constantly exchanging its genes throughout because, in a smaller population, new gene combinations and local mutations can become established more easily. And so, in such segments, small-scale changes in their gene structures bring about eventual mutual infertility; and this infertility defends their integrity, their separateness. That is the true origin of species. Darwin, unhelped by Mendel, never understood how this defense against swamping came about.

And so it is that daughter species can rapidly become adapted to different environments. There are some good cases in paleontology, when fossils are abundant, in which such off-budding species can be detected and in which a daughter species, having achieved a better general adaptation to the environment, ungratefully comes back and replaces the species of origin. Paleontologist Robert Bakker calls this "Lizzie Borden evolution."[5] How often, and where, this took place in our own family is something for the future to settle.

A Look Ahead

This book is a history, not a textbook, and I will not dwell on further complexities or genetic principles. The preceding is a simplification, and there are further nuances. In any case, the whole complexion of things changes and progresses. Prophecy is a tricky business, but it seems clear that molecular biology will play a bigger role, especially when its applications to evolution are

---

more variable and also unstable, splitting into several daughter species before settling down to a new stasis. This kind of history is much more difficult to observe in higher animals, because such abundant fossils are hardly ever obtained. Nevertheless, the question of "stasis" is much argued at points in our own evolution.

5. Here is a well-known bit of doggerel for anyone not familiar with this reputed parricide:

> *Lizzie Borden took an ax*
> *And gave her mother forty whacks;*
> *When she saw what she had done*
> *She gave her father forty-one.*

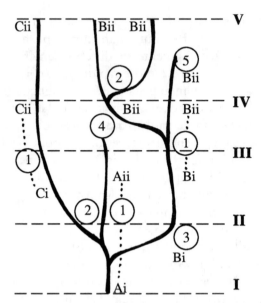

Fig. 2. A Crude Diagram of Some Kinds of Species Evolution

This shows five patterns:

1. Gradual change over time (Darwinian). Species A^i evolves to become A^{ii}; the same in B and C.

2. Gradual divergence (Darwinian). Species A^i divides to form A^{ii} and C^i.

3. Rapid change and/or migration (punctuated equilibria, or Punk Eek). A segment of A^i migrates to a new environment and by rapid response becomes a new species, B^i. The rapid change can also take place in situ.

4. Migration and rapid replacement (Lizzie Borden evolution). A segment of B^{ii} moves back into original home, where by better adaptation it replaces parental lineage, A^{ii}.

5. Extinction in one locality. For any reason (adverse conditions?) B^{ii} becomes extinct in its original home.

This also illustrates a problem naturalists have in dealing with species over time. There are two axes, horizontal and vertical. The diagram shows actual history; paleontologists usually see fossil exposures of one time level. Here, four time levels (I-IV) and the living present (V) are represented. Species must be recognized in two directions:

Horizontal — At present, or at given times in the past, related species, either in the same or different regions, can be told apart by comparing anatomy or, at present, by lack of interbreeding. The species problem is minor.

Vertical — A species changes continuously, as with A^i to A^{ii} or C^i to C^{ii}. There is no problem when differences are clear, especially with gaps in the fossil record, but if the record is continuous, it is an arbitrary matter when one species has become a different one.

better understood. And new techniques will turn up all kinds of fresh evidence — detection of microscopic scratches on fossil teeth to tell what food was being eaten is a small example.

Fossil discoveries will also become ever more plentiful. But there will be few blockbusters. Many finds will add gratifyingly to what we now see or tell us how to redirect our views, but there will hardly be wrenching revelations about the continent of our origins or the like. We are contentious as always but, we hope, on more and more solid ground.

Chapter 2

Patterns Without Plans

H ere is a paradox. First, look down over evolution's history from our place at the top. This shows us a pattern of life. We see many grades, from simple to complex, and fossils from the past reveal the advances from one level to another over time. Progress is not the only direction. Animals have branched out into every kind of life niche: gnats do things elephants cannot do, and vice versa. No wonder that it all seems like a plan, a grand design.

But now imagine looking forward from beginnings, with the eye, let us say, of an early primitive fish. That fish did not forecast what its myriad descendants would be like. There was then no

Kingdom	Animalia
Phylum	Vertebrata
Class	Mammalia
Order	Primates
Family	Hominidae
Genus	*Homo*
Species	*sapiens*

Fig. 3. How We Stand in the System of Classification
How living humanity is placed in the categories of life. Each category is part of the one above it. Classifiers freely use further subdivisions, like subspecies or superfamilies, as will be seen further on. The phylum is actually Chordata, to include a few oddities without spines, but with vertebrate organization.

blueprint to be followed, only unfoldings of opportunity. An environment is not hospitable all by itself: there must be life forms capable of using it. Air as a niche of opportunity was there, but the niche was empty until there could be insects, then birds and bats.

An environment can be no home at all, or it can be many different homes, depending on how many kinds of life have evolved to use them. In evolutionary terms, niches are really defined by the creatures inhabiting them. Look at what gathers at an African waterhole. Giraffes eat leaves; elephants eat whole branches; the hordes of antelopes feed on ground plants. Lions and leopards and hyenas can use the savanna only because their environment consists of so many animals to prey on. Contrariwise, because their particular environment is so shaped by lions, the antelopes have evolved their speed and alertness, and the buffalo their massive power. Australia, lacking such efficient carnivores, once had large, slow, amiable herbivores that could not have lasted a minute in Africa. We assume they were amiable: they appear to have become extinct at the hands of entering human hunters. Not having evolved a natural fear of predators, they would have permitted the dimmest of humans to walk up and stick a spear into them.

In evolution nothing is predetermined in detail but nothing is really accidental. There are simply rules of nature and life, like water running down hill. Water always does so, but what it runs over, or around, or through depends on a combination of that rule and a changeable landscape. Evolution has trends of adaptation, like that which produced one-toed horses who did better what three-toed horses were doing well. Such trends seem like a long groove of evolution. And there are what seem like flukes, changing the apparent direction of such a trend or starting a new one, suddenly opening new directions for adaptation.

Consider our vertical walking. Had I been given the charge of designing a human animal, I would have considered one with two arms and *four* legs, something like a centaur. We could run faster and stand still with less fatigue, and so on. But I was not consulted. Our first land ancestors needed no more than four limbs to scramble around on, and so they abandoned any hope of more appendages from still earlier ancestors. That is why land vertebrates (tetrapods, or "four legs") from frogs to gazelles have four limbs and why we have had to stand on our two hind limbs only, in order to rule the world with our forelimbs. But this last was not in the program originally. The earliest primates, our branch of mammals, did not forecast humanity — they only made it possible.

Survival and Extinction

We look around at the magnificent panoply of life today, every animal species with its badge of success. But numerous as present species may be, a far greater number of earlier species have gone to the trash heap of extinction. That is a major factor in whatever "plan" or "pattern" we may see at any time.

Of course there has been progress: the better tends to displace the good. We ourselves are nothing if not progressive. Think of some of the physical performances we are capable of: a concert pianist, a figure skater. These extraordinarily exact motions are made possible, ultimately, by the bony fin-stiffeners of a species of thick fish that began to slither around on land about a third of a billion years ago. Little by little, that fish species managed this ever better, as its muscle arrangements responded by natural selection to opportunities for more efficient slithering, and by the same process its bones responded to opportunities for better muscle attachments and body form.

Why Vertebrates Won Out

That is the story of vertebrates, or part of it. Most simply, it is expressed in the skeleton, which is what we can see in fossils. When one eats modern fish, one watches out for the bones, which are all ribs and backbone. But the ancient lobe-fin fish wore their fins on stumpy little lobes of flesh, and these contained small splints and slabs of bone, luckily numerous enough to become the multiple bones in later arms and legs. The amphibians, once on land, settled down to four

Pisces, the fishes
    Jawless (lamprey eel)
    Armored (totally extinct)
    Cartilaginous (sharks)
    Bony fishes
        Lobe-fins
        Ray-fins
Tetrapoda, the land vertebrates
    Amphibians
    Reptiles
    Birds
    Mammals

The Classes of Vertebrates

limbs, each with five toes, all supported by real limb skeletons. The forelimbs were hitched to the skull, but reptiles, evolving from the amphibians, developed distinct shoulder and hip girdles, with vastly better engineering of the whole skeleton. Finally, the mammals simplified, strengthened, and improved these possibilities.

Those early amphibians, like their living descendants the salamanders, got around partly by pushing with their little legs and partly by bending their bodies from side to side. Fish swim with the same side-to-side motion, and the tailfins of fishes are vertical. Compare a running greyhound, whose legs virtually dominate his skeleton, and whose spine bends and unbends vertically to a phenomenal degree. And such mammals as have gone back to the sea, like whales and dolphins, also bend vertically, not side to side, and their flukes are horizontal. A very different kettle of fish.

Our limbs likewise dominate our spines, which are very flexible in the neck and the lumbar part. As things go, we are an extremely new species of animal and, of course, a unique one. But the fundamentals that made us possible, the basic patents so to speak, are old. We are, first of all, vertebrates and, second of all, mammals.

When we think of "animals" we are apt to be thinking of vertebrates, not clams or worms or insects. Vertebrates rule the land, sea, and air because they are highly mobile, have excellent senses, and can be large-bodied. They have internal skeletons with a spine made up of separate vertebrae, and (with exceptions like fish, whales, and snakes) they have pairs of limbs. Insects and lobsters likewise have legs and a head end, but their skeletons are external, an arrangement that does not work out well for large body size. A growing lobster must repeatedly shed its shell and form a new one.

Worse, insects must get their oxygen through little ducts from the outside, and air cannot penetrate very far in this scheme. But vertebrates efficiently extract oxygen from the air or water using lungs or gills. They transport the oxygen all through the body by a blood stream, so the size problem is solved. The heart can drive blood all the way up a giraffe's neck or throughout a blue whale, which is quite a trip. And vertebrates have a unified nervous system and a brain, which organizes the senses and muscles, as well as things like temperature and hormones, to control the whole animal efficiently. In general, the secret of vertebrates lies in the special central systems they have evolved.

As you go up the scale toward the mammals, organization becomes more complex and brains get larger. The process all goes together. The vertebrates, with all their potentials for perception and action, give brains more scope. Brains did not evolve without good reason; for example, worms and snails lack the senses or the behavior to make many demands on a nervous organization. But we are mammals, who indeed have uses for complex brains and nervous systems. In the end, evolution simply tends to produce what can be put to use.

Getting to Land

The history of all this is pretty well known, and becoming more so. Signs of life — one-celled creatures — seem to appear in deposits (e.g., in Australia and Greenland) laid down more than three billion years ago. (This in itself — creatures with a cell wall and some internal organization — was already a key achievement in evolution.) Much later, many forms of soft-bodied creatures developed and have been found in a few places where conditions for preserving their impressions were exceptionally good. But only in the Cambrian period, beginning about 590 million years ago, did there appear animals with enough hard parts to make plentiful and good fossils. All water-dwellers, they were in fact so varied that paleontologists talk about the Cambrian "explosion." Many of the animals are recognizable as belonging to groups still present; some others seem to have had kinds of organization that no longer exist. Extinction never sleeps.

Not long afterward — a little under half a billion years ago — very primitive vertebrates appeared. They are hardly recognizable as such; they lacked jaws or limbs and had a barely detectable skeleton; the typical ones were somewhat fish-like bottom-grubbers. They were followed by other major classes of "fishes." The fish we know today are the most numerous of vertebrates by all odds: the number of living species is vast, and we eat fish by the billions every year. This great group is called the "ray-fins" because of their light, strong fins ribbed by spines.

These familiar fish are actually a rather late development, becoming the dominant form of fish life a mere 100 million years ago, as though evolution took its time deciding what a fish should really be like. And they are not the ancestors of land vertebrates. A second non-ancestor is one of the other earlier classes of fishes, the sharks. They are different in many ways, especially in having skeletons that remain cartilage, bone's precursor, and do not turn to bone as in other vertebrates. Greatly varied, sharks were, and are, perfect animals in their own way, having been successful and little changed for hundreds of millions of years. No modern bony fish, from minnow to tuna, grows to anything like the size of the larger sharks.

Another early and successful branch of the fish-like animals was that of the lobe-fins, close to the ray-fins in origin. Their species were numerous also, living mainly in inland freshwater seas. Their stumpy little "limbs" with bones in them have been described. In addition, the lobe-fins had small internal air sacs that could pass oxygen from swallowed air to the blood. (The ray-fins also have the sac, but can now use it only as a ballast tank, to regulate buoyancy.)[1] Lobe-fins of one or another line were the ancestors of the land vertebrates.

But they themselves became virtually extinct. Only fifty years have passed since it was first discovered that one or two species of lobe-fins (of the group known as coelacanths) still exist off the eastern coast of Africa, mostly around the Comoro Islands. This is probably an example of chance survival by adaptation to an environment (the saltwater ocean) different from that of the main group.[2] Other surviving relatives of the lobe-fins are the lungfish of Australia, Africa, and South America. These manage to live through the dry season by holing up in the caked mud of their lakes and breathing barely enough air to stay alive. It seems likely that some branch of the widespread lobe-fins faced the same predicament but, instead of cravenly subsiding into the mud, they gamely began to move about, bending from side to side, pulling themselves along

---

1. Both the air sac and the limb skeleton are very good examples of an evolutionary phenomenon usually called "pre-adaptation." This means that some element, either one still performing a function or one that has lost its function, is taken over and put to a new use. A striking case is the mammal ear. Reptile lower jaws are formed of several bones involved in the joint with the skull. In mammals a new, stronger joint formed, and the lower jaw is a single bone. Instead of disappearing, the other elements became the delicate bones in the mammalian middle ear.

2. It is not known why these particular coelacanths survived, or why so many kinds of lobe-fins became extinct. The living ones move their lobe-fins on alternate sides, like land animals, but for swimming, not for bottom walking. These coelacanths are not believed to be land animals' closest relatives among the lobe-finned fishes, but molecular tests show that they are indeed relatives. With their rarity and their scientific interest they may well become extinct under our noses, like the dodo, the passenger pigeon, and the Tasmanian wolf.

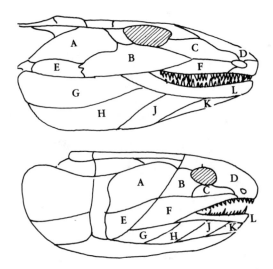

Fig. 4. Skull Bones in a Lobe-fin Fish (below) and an Early Amphibian (above)
Letters mark the same bones in the two skulls, demonstrating in part the correspondence in the relations among many of the same bones.

with their primitive legs, looking for a better pond. And some new opportunities — perhaps food in the form of stranded fish — showed that this was a paying proposition. In any case, oxygen starvation probably played a major role. No better example could be found of the role of happenstance in evolution.

So the lobe-fins almost vanished. In the water, their ray-fin cousins were eventually more successful but, once ashore, the lobe-fins gave rise to the first amphibians, the original land vertebrates. These crossed the Rubicon, living on land and going to the water for food. The transition is clear: in fossils of the lobe-fins and the first amphibians, details of skeleton and skull are very similar, including the reshaped bones in the new, small limbs.

Chapter 3

Reptiles and Mammals

Whatever the blessings of life on land, the new amphibians had severe problems to face as well. Their lungs were extremely primitive, and modern amphibians have to absorb oxygen any way they can: through the wet skin or by gulping air with the mouth. And they were, as they still are, normally tied to water, by their water-laid eggs and by having to live through the tadpole stage.

The Economy of Reproduction

Typical fishes lay great numbers of eggs directly into the water, with the male spreading milt to fertilize them. Not many of the little ones get far in life; most go to feed other fish. Higher vertebrates are more economical and successful with their offspring. Amphibians are less prodigal with their eggs than are fishes, although a tadpole's chance is still pretty slim. Reptiles scored a vital advance by wrapping the tadpole stage up in a large egg with a shell, containing membranes through which oxygen can pass, and also containing enough food for development into a small but full-fledged reptile. Birds, which are follow-up reptiles, have gone a step further by more care of eggs and young. So these last two classes enjoy much greater economy and efficiency in reproducing.

Finally, mammals have put the whole egg stage into the mother's body. Through some membranes left over from the reptile egg, the developing infant is fed from its mother's blood stream until it is a fair-sized copy of its parents. (This is the story in placental mammals; marsupial mammals like kangaroos or opossums put the fetal young into a pouch.) After birth the infant lives on mother's milk from mammary glands and is protected for months or years by powerful mother love. The slow development before and after birth allows time for the development of a complex nervous system and for learning to use it. So mammals keep the species going with a lot

Vertebrates	*Fishes, Amphibians, Reptiles, Birds, Mammals* Bony skeleton, jointed spine. Central systems for oxygenating and circulating blood, and for nerve sensation and muscle control.
Tetrapods	*Land Vertebrates: Amphibians, Reptiles, Birds, Mammals* Paired limbs with five or fewer digits: air-breathing.
Amniotes	*Reptiles, Birds, Mammals* Well-developed skeletons; fetal development in egg or womb with special protective membranes like the amnion to supply food, oxygen.
Mammals	*Milk-producing Vertebrates* Warm blood with mechanisms for temperature control (fur, fat, sweat glands, etc.); two sets of specialized teeth; live birth and parental care.
Primates	*Lemurs, Monkeys, Apes, Men* Nails, not claws; opposable thumb; skin friction pads or patterns; color and stereoscopic vision.
Hominoidea	*Apes, Men* Large size, tailless, broad molar teeth with common pattern, large brains.
Hominidae	*Present humanity back to post-ape ancestors* Erect walkers, smaller canine teeth.
Homo	*Hominids of the last 2.5 million years* More modern skeletons, larger brains, tool-makers.

Fig. 5. Where We Stand Among the Vertebrates
A Russian-doll arrangement showing the main characters we get by virtue
of belonging to successive subdivisions of the vertebrates.

of trouble but with little wastage. Few young are born at a time;
many species, like ourselves, as a rule bear only one. That is enough;
reproductively, we are appallingly successful.

To go back a few hundred million years. The early amphibians
were quite successful and of fairly good size: they were several feet
long and looked like a hastily designed crocodile. Later, but still over
three hundred million years ago, the first reptiles arose, with a lighter
skeleton and better limbs. In addition they had that key advantage,
the protected egg, which freed them from having to start life in the

water. (Reptiles, birds, and mammals are grouped together as "amni-otes" because they all grow a surrounding fetal membrane, the amnion, whether in an eggshell or a womb.) Some reptiles, like turtles and alligators, do very well in water; others, like snakes and lizards, do very well in deserts. In the face of these advanced models, the central core of the amphibians collapsed, leaving essentially only frogs and newts, barely enough to keep a witch in brew.

Far back in reptile history, ancestors of mammals came on stage, followed by mammals themselves. Eventually these last took over the earth; once more, the center of the old order gave way, leaving surviving reptiles only in niches on the fringe where mammal competition was absent: as above, snakes and turtles. But the story is not so simple and seems to have been punctuated by a vast catastrophe.

The Age of Reptiles

Coming into their own, for a hundred million years the reptiles were a magnificent success. No wonder troops of small children go to the museums and crane their necks at the monstrous dinosaur skeletons. Not all dinosaurs were large, and not all large reptiles were dinosaurs, which were only one main division of reptiles. The impressive thing is how reptiles evolved in so many directions, taking up nature's niches very fully. Surely, people who give their lives to the study of dinosaurs are happy people.

Some dinosaurs, such as tyrannosaurs, ran on their hind legs, as bipeds with big tails. In their later evolutionary radiation, except for possible extinct oddities, the mammals never produced animals like this — what kangaroos do is hop, not run, and we ourselves, with our straightened-out legs, are something else again. (A few living reptiles can run on two legs; and of course birds, which descended from dinosaurs, got their start as bipeds.)

Reptiles re-entered the sea in various forms, including the dolphin-like ichthyosaur. The flying pterodactyls were no isolated oddity; pterosaurs lasted for millions of years and fanned out in many species: some were only as big as a robin, but one species had a wing spread of nearly forty feet. No flying bird has ever approached that size. The wings of pterosaurs were different in structure from those of both birds and bats and it is thought they had evolved a skeletal forelimb shape that allowed a kind of wing stroke that birds

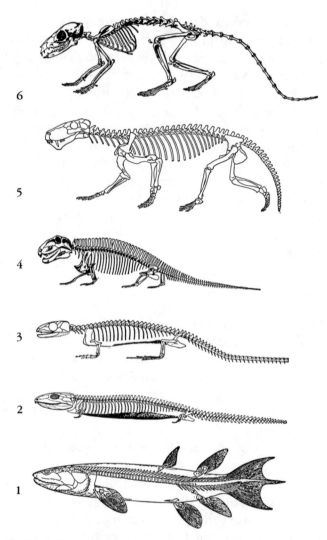

6

5

4

3

2

1

Fig. 6. Skeletal Changes Leading to Mammals
In this progress, the limbs become more important and better articulated
with the spine; the spine becomes freer and more mobile, without ribs in
the neck and lumbar regions; and the ribcage becomes more of a special,
organ-containing unit. From the top down:
6. A generalized placental mammal
5. A typical mammal-like reptile
4. A reptile ancestral to mammal-like reptiles
3. An early and primitive reptile
2. An early amphibian
1. Lobe-finned fish

never achieved. Their skeletons were extremely light. There is a lot still to be learned about pterosaurs.

One early reptile group was that of the mammal-like reptiles, who were close to the original reptile stem and remained four-footed, being progressive and generalized at the same time, and they bequeathed this basic ancestral form to their mammal descendants. They anticipated mammals in a few ways; for example, some are suspected to have been warm-blooded because their fossils have a bony palate to separate the mouth and nose cavities, an arrangement important for the constant breathing necessary to maintain body warmth. Interestingly, the mammal-like reptiles did not produce much of a radiation, and eventually declined and vanished from the scene. But, and still in the time of the reptiles' early expansion, true mammals had also arisen out of the mammal-like reptiles. So they were present through much of this time but remained in the background and, compared with the dinosaurs, they were small, few, and somewhat nondescript. Then, at the end of the Mesozoic era, sixty-five million years ago, the great mass of reptiles disappeared.

Episodes of widespread extinction, followed by expansion of surviving forms, have occurred from time to time throughout the past. Two such episodes have been especially severe: one marking the transition from Paleozoic (Primary) life forms to those of the Mesozoic (Secondary), and one marking the transition from the Mesozoic (Age of Reptiles) to the Cenozoic or Tertiary (the Age of Mammals). Both transitions witnessed huge geological catastrophes. The first, dated at just about 249 million years B.P., was marked by enormous lava flows in Siberia. In the second, sixty-five million years ago, an asteroid, or pieces of one, struck the earth, evidently in northern Yucatán. Suggested some years ago, evidence for this has been steadily accumulating, particularly in the form of special minerals most likely to have come from space, deposited just at the boundary layer in many parts of the world.

Exactly what happened is a matter of lively interest, along with speculation as to an ensuing period of very damaged climate and loss of vegetation. Final development of this history is something for the future, and details do not matter much here. Great numbers of species of vegetable life and ocean plankton disappeared, along with many animal forms. The dinosaurs were the most spectacular to go, from our point of view. They are traceable in the rocks right up to the boundary level and then they vanish.

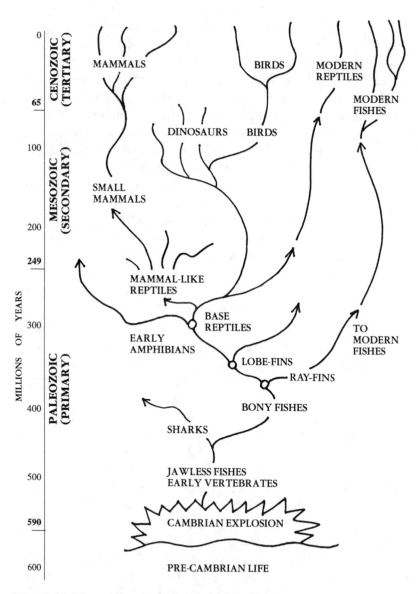

Fig. 7. A Time Chart of Vertebrate Evolution
This locates the general times of emergence of main groups of vertebrates.
Especially important are the nodes of branching of ray-finned and lobe-
finned fishes and of the basic reptiles. Like the dinosaurs later, the early
mammal-like reptiles were numerous, but eventually declined to extinction, leaving only a minor number of actual mammal ancestors.

So the dominion of reptiles ended, leaving as in the earlier case of the amphibians only fringe survivors: snakes, lizards, turtles, and crocodiles. The mammals of that time were small and simple and had not been able to punch their way through the ruling reptiles to any prominence. Without question, mammals are more highly evolved, with many significant advantages. But it is not at all obvious that it was these advantages that enabled them to overthrow the reptiles; there was no David-and-Goliath story. Rather, the dinosaurs' fate was the mammals' good fortune, simply because the mammals survived, as very small beasts of no great account but without the special needs of existing dinosaurs. Being mammals was less important than having a clear field for an evolutionary radiation, as the reptiles had before them. Sixty-five million years ago, as the Tertiary era opened, that radiation began.

The Deployment of Mammals

Simple or not, the mammals came into their own with a set of clear basic improvements. Many of these come together under the heading of "warm blood," meaning a constant temperature near 98.6 degrees. This gives us all high activity of body and brain whatever the weather, although we pay the price of needing a lot of food and a lot of sleep. Our steady body heat is maintained by these things: fur or hair to keep us warm, and sweat glands to keep us from getting too warm; complicated teeth for efficient chewing; better arrangements of the blood stream as it passes through heart and lungs; better senses and brain, and so on, including of course live birth and maternal care.

Live birth
  Mammary glands
Warm blood
  Hair and fur
  Four-chambered heart
  Two sets of teeth
  Varied tooth shapes
Flexible skeleton
  Growth at epiphyses
  7 vertebrae in neck
  Free lumbar spine
  Lower jaw a single bone
  3 bones in inner ear

Marks of Mammals

Warm blood seems to have been an idea whose time had come. There are signs of its appearance in the mammal-like reptiles and probably in at least some dinosaurs. And like early automobiles, some of which ran on steam or electricity, mammals appeared as several early stems. Besides our own division, that of the placental mammals, only two such stems survive: the many marsupials of Australia (and a few elsewhere) and that peculiar pair of egg-layers, also Australian: the platypus and the spiny anteater, who nevertheless nurse their young. But like gasoline-powered autos, the placental mammals simply overran the competition and took command of the field wherever they arrived.

The New Weapon

Two mammalian advantages should get special mention. The first is teeth. Other vertebrates make do with simple points or pegs (sharks, for example, do very well). Reptile teeth may have a little modeling or extra cusps. There may be numerous teeth in the row, and they fall out and get replaced by new ones. Sharks lose and replace teeth by the thousands.

Mammals have two sets only, one to grow on (the milk set) and one to live with. But the teeth have hard enamel crowns, which are complexly sculpted to serve various functions. A smiling crocodile shows peg teeth all the way. A snarling mammal shows incisors in the front for nipping, a canine or eye-tooth at the corner for gripping or slashing, premolars (bicuspids in ourselves) for cutting, and molars or grinders at the rear. Our own pattern, though clear, is rather uncomplicated; other kinds of mammal have met their special

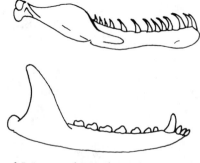

Fig. 8. Reptile and Mammal Teeth
Lower jaws of a snake and a simple mammal. The reptile has undifferentiated pointed teeth. The mammal has 3 incisors, 1 canine, 4 premolars, and 3 molars (in each half jaw). We ourselves have lost 1 incisor and 2 premolars (bicuspids).

needs with special shapes or changes. Carnivores have made a meat-cutting row out of their cheek teeth. Elephants have made tusks out of a couple of incisors and complex millstones out of the molars. And so on.

The advantage gained, of course, is efficient chewing and quick digestion, to support that expensive metabolism of warm blood. Carnivores bite the meat off the bones. Squirrels bite the shells off nuts. A snake or a lizard cannot do this; their prey is swallowed whole and the swallower (say, a boa with a pig) may be out of circulation until the meal has been digested.

In any case, mammal teeth are immensely varied in details of shape. And that is the best basis on which a paleontologist can recognize the mammal species whose bones he has found in the ground. Or, if he does not recognize the species, he can name a new one. In that case, of course, he had better know all the old ones.

The Elegant Skeleton

A second new mammal improvement lies in the skeleton. Reptile bones simply grow at both ends. Mammal bones early develop a separate piece of bone (an epiphysis) at a joint surface and some other places. This gives a firmer, well-shaped surface at a joint, while the bone goes on growing in length in the cartilage between the epiphysis and the shaft until growth is finished (in us, when a teen becomes an adult), and the shaft and the ends unite.

Beyond this, mammal skeletons throughout are better engineered for special functions than are those of non-mammals. It is as though mammals finished what reptiles had only started. Major parts are more distinct: neck, shoulder girdle, chest, lumbar spine, pelvis. Above all, limbs are greatly varied — look at horses, bats, and people. However, while these parts take highly special shapes from one kind of mammal to another, the basic plan remains the same in all. There is no difficulty in spotting the same elements in the three species just named. And every last species of mammal (there are only a couple of exceptions) has exactly seven vertebrae in its neck, from giraffe to dolphin, from mouse to mammoth. Every last *species*, that is, not every individual: you have probably heard of people who have only six neck bones. Genetic variation, as always.

Mammal skulls are simplified, with far fewer bones than in the ancestral fish, and are shaped very much according to eating habits. Anteaters are snouty, whales are snoutless. And in all mammals the lower jaw is a single bone, not a composite as in lower vertebrates; this makes it a more solid chewing machine.

The liberation of the mammals did not make human beings inevitable. But, by happenstance, the way was left open for us. Elsewhere doors of flexibility closed, as the mammals, like the reptiles before them, diversified into a number of main groups called orders. Many orders bought success at the price of drastic modification or loss. Modern horses, superb large runners, have only one toe on each foot. Whales have no hind limbs, only funny little bony remnants of a pelvis; and many whale species have lost all their teeth. Take bats; we seldom see them but they are there, please be sure. Their arms have become skin-covered wings. Bats own the night air, after people and birds have gone to bed. Feeding on mosquitoes, fruit, or blood, they have been hugely successful, with a greater number of species than most mammal orders and colossal crowds of individuals. But, in terms of evolution, bats became bats at least fifty million years ago, and that was the end of the road. Our own order did not get into such a closed alley of adaptation.

Chapter 4

The Primates

We belong to the order of the Primates, and it is hard to see how any of the other orders might have given rise to ourselves. While most mammal orders have moved out on specialized vectors of adaptation, the primates have departed less from an original mammal form; they are generalized, conservative, and, in a sense, primitive. But they have explored the possibilities of improving on that general form rather than limiting it by going in for wings or hooves.

Fortunately, we can see the stages of improvement, because the living primates are spread out over those stages for our inspection. Today's primates are a broad group: the less progressive kinds have been left off at various evolutionary way stations; they are not actually living fossils, since they show their own gentle progress, but, nevertheless, they represent older primate levels. So, while we ourselves crown the primate order by reason of our unparalleled status, there are also the very intelligent anthropoid apes, all the monkeys, and the various kinds of lemurs and lorises. These last two, which are the lower primates (called the "prosimians"), are separated as a suborder from the higher primates (called "simians" or "anthropoids").

Good Things About Primates

Because they have stayed closer than most orders to the original mammalian ground plan, it is less easy to point to well-marked distinctions of primates. It has often been said, and aptly, that they are distinguished less by clear anatomical badges — like a bat's wings or a rodent's ever-growing front teeth — than by certain trends of evolution. And these trends are indeed progressive.

Nevertheless, all the primates have some things in common. Very important is the power to grasp things or hold onto things with the fingers wrapped around one side and the thumb on the other side. The thumb is opposable; that is, it can be rotated to face the other digits. One piece of good fortune is that primates kept all five digits from the early amphibian hand and foot. To these endowments are added friction skin (fingerprints) and nails instead of claws on each finger and toe. Thus, primates clasp things instead of digging into them with claws.

Power of grasp
  Five digits on hands and feet
  Opposable thumb
  Nails in place of claws
  Mobility of arms and legs
  Well-developed collar bone
  Friction skin
Development of vision
  Stereoscopic vision
  Full color vision
  Complete ring around eye socket
  Eyes move toward front
  Face shorter, more vertical

Marks of Primates

Together with this, primates also have an increased mobility of the original bones of the lower arm. Hold your right elbow with your left hand and see how easily you can turn your right hand through 180 degrees, without moving your elbow at all, as your radius (on the thumb side) rotates on your ulna. Also, you can feed yourself raspberries by reaching out your hand almost three feet to pick them off the bush and then folding your elbow to put them in your mouth without otherwise moving. With either arm you can touch almost any part of your body, except the same arm. This great flexibility is just what horses do not want; horses have now evolved their digits down to a single large toe while also losing the independence of the radius and ulna. (The earliest known horse ancestor had four toes on its forefeet.) Like horses, we ourselves have evolved somewhat more stable feet; other primates have more arm-like hind legs.

This power of flexibility and grasp evidently progressed because of the primates' use of trees as a major part of their behavior. Human beings are the only earth-bound primates. Gorillas, with their great weight, also spend most of their time on terra firma but are agile in trees all the same, and young gorillas make their sleeping nests aloft.

Also related to tree life is another property of primates: acute vision. This, together with eye-hand coordination, gets better in the higher primates. They have full color vision which is also stereoscopic. That is to say, there is full overlap of the fields of vision of both eyes, and merging of their nerve endings in the brain allows a close estimate of distances. It is easy to see the usefulness of this acuteness of sight for jumping from one branch or tree to another, or for surveying flowers and fruits with a view to consumption. The increasing overlap of fields of vision in primate evolution is expressed in the way the face becomes less snouty and the eyes move around to the front as we go up the primate ladder.

These changes also express another primate distinction, the decline in the sense of smell. It is obvious how important this sense is to mammals living on the two-dimensional surface of the ground. We see them sniffing constantly and smelling things we take no notice of. A prey animal can hide itself from sight and sound by concealment, by camouflage, and by keeping motionless. The one thing it cannot do is stop smelling of whatever it smells of.

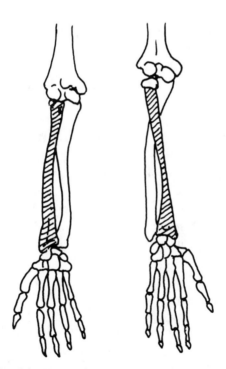

Fig. 9. The Flexible Forearm
How, in the right arm, the radius (shaded) rotates, at its upper end against the ulna and the humerus (upper arm), to turn the hand over. Its lower end is on the thumb side. *Left*, palm of the hand faces forward; *right*, it faces back. The humerus and ulna do not move.

Smell is far less valuable in the three-dimensional world of the trees. In primates the smell part of the brain is de-emphasized, and so is its instrument, the snout. Thus, the eyes move around into the frontal plane, increasing stereoscopic vision, and bit by bit, the bony eye socket in the skull gets a complete wall at the sides and back. The nose becomes shorter and narrower, the face and mouth drop down, and the face becomes more vertical.

So does the body. Cats and dogs, squirrels and rabbits, may all squat in a semi-erect position when at rest. Likewise most primates; but erectness is more usual and habitual among them since they practice it when climbing or swinging in trees. Uprightness is most marked in our nearest relatives, the apes, as they move by swinging from branch to branch; and the head is perched as much on top of the neck as in front of it; the latter is its position in a horizontal mammal. So our own erectness, a hallmark of the species, is something thoroughly prepared for in our ape cousins.

These are some of the tendencies or trends in the evolution of primates. Further trends are toward larger body size and larger brains; but various other orders of mammals have shown these trends, which are apt to be advantageous. I am not saying that we have simply gone along with mammals in general, because of course we, and the apes and monkeys also, have outshone the rest intellectually. And our nimble arms and hands, together with our ability to see very precisely what we are doing with them, have certainly been moving forces in brain evolution. But it is well to realize that brain power is a very general kind of advantage among mammals (as it apparently was to some of the later dinosaurs before them), always within the limits of a species' capacity to make use of it. A three-toed sloth, forever hanging from a branch, has all the brain he needs, which is not much.

The Living Primates

A catalogue of primates starts with the lemurs. Today there are many and varied species, some tiny like the mouse lemur but most about the size of a fox. They tend to be attractively furred and colored. The most common, the ring-tailed lemurs, run about on all fours with tail straight up in the air, while other species, like the sifakas, hop along upright, very smartly, on the ground, and also like to sit upright in trees, nestled in the fork of a limb.

PRIMATES
Prosimii
Lemurs (Madagascar)
Lorises (Africa, Asia)
Tarsius (Southeast Asia)

Anthropoidea
Platyrrhini (New World monkeys)
Callitrichidae (marmosets)
Cebidae (other monkeys)

Catarrhini (Old World higher primates)
Cercopithecoidea (Old World monkeys)
Cercopithecidae (macaques, baboons)
Colobidae (mainly Asia: langurs, colobs)
Hominoidea
Hylobatidae (gibbons)
Pongidae (great apes)
Hominidae (all human beings and australopithecines)

Fig. 10. A Classification of Living Primates
This is meant to show the divisions of primates as dealt with herein. It is not fully formal, and classifications differ in detail among authorities; also, names used differ from one to another. In one important matter, some writers divide *Pongidae*, reserving this family for the orang utan alone and placing gorilla and chimpanzee in *Hominidae*, for reasons discussed in the text.

All the lemurs live on the island of Madagascar, off Africa's east coast. Somehow, one or two species got there in the distant past, most likely by rafting. Today, rafts of fallen vegetation get carried out to sea by the Limpopo and Zambesi rivers, and animal castaways are often seen on them. For millions of years Madagascar was a paradise for lemurs. Facing no competition from predators or more advanced primates, they expanded into the many lemur species of today and into others now extinct. In recent times there existed a number of very large forms, with one, *Archaeoindris*, estimated to have weighed over 400 pounds. That is gorilla territory.

The big lemurs disappeared in the last couple of thousand years. No one knows why, but the first human beings arrived on the island about two thousand years ago, and that was probably more than a coincidence.[1] Discovering or rediscovering an occasional new species and finding remains of others now extinct, devoted lemurologists have been having a field day in Madagascar, and no wonder: the island is a splendid evolutionary laboratory.

Prosimian relatives on the continents of the Old World have not been so fortunate: only a sprinkling of these cousins of the lemurs has survived. The lorises and bushbabies of Africa and Asia have had to face the harsher world, surely because graduate-school primates like the monkeys can do almost anything prosimians can do and do it better. Little is known of loris history, but they survive as night-prowlers, living mostly on insects and very small game. The bushbabies of Africa are efficient springers and jumpers, with elongated feet. The lorises are slender and long-bodied, long-limbed, and short-tailed, quite the opposite of the bushbabies; their eyes are big and sad, and they move so slowly before making a final grab at a locust that it is positively annoying to watch them.

The spectral tarsier of Southeast Asia fills the niche there of the bushbabies of Africa. He is quite distinct in various ways from all the lemurs and lorises, noticeably in having a nose of plain skin like the higher primates rather than the moist muzzle of most mammals, including prosimians. Thus, he has often been classified as part way between lemurs and monkeys. He has a midfoot (the tarsus, hence his name) much more elongated than the bushbaby's, so that he springs and jumps from one tree or branch to another in an upright position. He is only rat-sized, but his eyes are enormous, bulging from his skull, giving him splendid night vision. He is more than a curiosity; he might be an important signpost in our own evolution. This is one of those problems in our past.

---

1. New Zealand, also isolated until the arrival less than a thousand years ago of the Polynesian Maoris, had a number of species, large to huge, of the flightless ostrich-like moas. The Maoris killed them all in due course, missing only their tiny relative, the furtive night-hunting kiwi. For other living things, the coming of human beings often makes an excellent example of a "catastrophe" in evolution, leading to still another "extinction event." Certainly, such a manmade catastrophe is in full swing worldwide at the moment.

The Primate Elite

The higher primates, sometimes called "simians," are formally named the Anthropoidea. They are respectable denizens of the day (as are the lemurs) and have avoided certain specializations seen in the prosimians. Their teeth, for example, do not display various reductions and modifications of shape found in the latter. The original set of mammal teeth is believed to have consisted of three incisors, one canine, four premolars, and three molars in each half jaw, represented by the formula of 3–1–4–3. All higher primates have lost one incisor and one premolar, and this is what is seen in the South American monkeys: 2–1–3–3.

Many, but not all, of these American monkeys can hang by their prehensile tails. They also seem to have better hands on their legs than on their arms. A howling monkey's feet are good graspers, but his hands are not very skillful; the New World monkeys appear to be adapted for a secure grasp at the rear while exploring and feeding at the front. All this is fitting in these highly arboreal animals who avoid the forest floor.

Among the higher primates, the American monkeys are set off in many ways from those of Africa and Asia, ways in which the latter are more closely related to ourselves, as Darwin knew. For an obvious thing, we all have the same tooth count, with one fewer premolar

Fig. 11. Noses in Platyrrhines (left)and Catarrhines (right)
The New World (*left*) and Old World (*right*) monkeys are similarly monkey-like in general facial appearance, except for the wider septum between the nostrils in the New World monkeys.

than American monkeys, making our formula 2–1–2–3. There are various particulars of bones in the skull that the Old World monkeys share with the apes and ourselves, and the two nostrils of the nose are close together, separated only by a narrow septum, unlike the more widely spaced nostrils of the American monkeys. From this last feature, these New World primates are called the *platyrrhines* (flat-nosed), and we call all the Old World forms, ourselves included, the *catarrhine* primates (downward-facing nose). This also lets you know that a Tarzan movie has been made on the cheap when you see the hero parading around "Africa" with an American monkey on his shoulder.

The Old World monkeys, although highly varied in appearance, are probably mutually more closely related in ancestry than are the many American forms. The two great tribes of African baboons and Asiatic macaques are essentially ground-dwellers, while of course being at home in trees for purposes of refuge. The macaques are adaptable, with species ranging from India and Indonesia to colder Japan and North China. In late Tertiary time they extended right across Europe, where they now survive only on the Rock of Gibraltar as the "Barbary Apes." The other Asian monkeys are mostly slender, leaf-eating langurs, but the large nose monkey of Borneo is both beautiful and bizarre, with his handsome gray and red coat and his pendulous fleshy nose. The monkeys of Africa are similarly decorative and too varied to catalogue here. They are well worth the trip to a major zoo.

One thing common to all the Old World monkeys is bilophodonty. This means simply that their cheek teeth (the premolars and molars) have evolved crosswise crests, two in the case of each molar, connecting the two forward and two rear cusps, respectively. Bilophodonty is supposedly efficient for chewing vegetable matter. In any case, it has evidently evolved away from a more standard and ancient arrangement of the molar cusps, which is the one we have in our mouths and the apes in theirs.

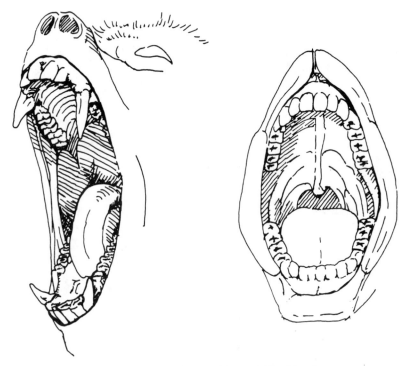

Fig. 12. Wider, Please: A Baboon Shows His Teeth

Four things about a baboon's teeth (*left*) may be noted:

1. The three molar teeth are bilophodont: they each have two crossing crests of two cusps, which marks them as teeth of Old World monkeys.

2. The first lower premolar — behind the canine — has a cutting edge where it shears against the upper canine; in ourselves both premolars have the same bicuspid shape.

3. The dental formula is 2–1–2–3, meaning that, on each side, there are two incisors, one canine, two premolars or bicuspids, and three molars, which marks the baboon as a catarrhine primate, like ourselves.

4. The canines are very large, fitting into gaps in the opposite jaw, with the lower canine biting in front of the upper. Although human canines do not protrude, or do so only slightly, their position is the same, with the tips of the lowers being in front of those of the uppers.

Human beings cannot gape in this way (unless the cheeks are sectioned as shown at *right*) but the comparison shows the de-emphasized front teeth and the simpler, non-bilophodont teeth of our set.

Chapter 5

The All Important Hominoids

Now we come to an interesting group of animals: the hominoids. They are so named because they make up the superfamily Hominoidea, and they differ in several ways from the Old World monkeys (who make up the superfamily Cercopithecoidea). Instead of bilophodont teeth, the hominoids have an older, less special molar pattern; they have broad shoulders and flattish chests; and they have no tails. Today's hominoids are the gibbon, the three great apes, and ourselves.

The gibbons of Southeast Asia are on a limb by themselves. They are the smallest apes, and are furry rather than hairy like the rest of us. They have extremely long arms: fabulous aerial acrobats, they swing rapidly along from limb to limb, arm-leaping across wide spaces, often twenty feet wide. Their hands are like flexible hooks, not grasping a bough as they hang and swing but merely hooking as they go, without using their thumbs. A gibbon is almost never on the ground; when he is, he runs along with his arms out, as though doing a balancing act. He does the same thing walking along a limb.

But gibbons are not our close relatives. It has become increasingly clear from their fossil genealogy that they evolved separately over many millions of years. Like the other apes, they arrived at a body form adapted for brachiating (swinging through the trees by the arms), but they evidently reached this form independently. In fact, the gibbons have a set of chromosomes more like that of monkeys than like the big apes. Nevertheless, other molecular studies, as well as much anatomy, show gibbons to be members in good standing of the hominoid clan.

The Anthropoid Apes

With the great apes, however, we are face to face with beings of crucial importance to our knowledge of human evolution. They are,

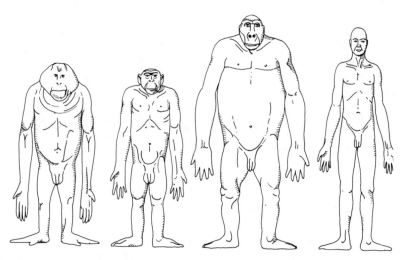

Fig. 13. Body Proportions in the Great Apes and Man
Presented here are average body outlines of the large apes and a human,
shown without hair. Limbs and body posture, especially of the apes, have
been twisted into unnatural positions in order to show relative lengths
better. *Left to right*, orang utan, chimpanzee, gorilla, and man.

of course, the orang utans of Borneo and Sumatra and the chimpanzees and gorillas of the African forests. We are much closer kin to
them than to any other animal, and vice versa. They do not play
chess, practice law, or pollute. But in physique, brain form, and body
chemistry, as well as in much capacity for behavior, they are over-
whelmingly similar to us, and we need to know all about them in
any study of our own evolution. Before getting down to the essentials
of likenesses and differences, let us look at the family album.

Clearly the closest to us are the two African apes. These are
essentially ground-dwellers. Normally they walk on all fours, using
the flats of their hind feet, but with the great toe separated like a
thumb, while they lean on the extended knuckles of the hand, so
that they are semi-erect as they go. Of course, they are adept in trees,
and at night chimpanzees sleep in tree nests made by folding
branches together. Young gorillas do the same, but older ones make
their nests on the ground.

The main gorilla habitat, in the volcano region of East Africa,
is becoming cramped as the cattle of this overpopulated region are
driven higher and higher up the mountains for grazing. Worse,

poachers think nothing of shooting a whole gorilla family to kidnap a baby for sale. So the gorillas are a highly endangered species.

A gorilla is something like a scaled-up version of a chimpanzee. The two look different, the chimp with his jug-handle ears and his brown or mottled skin, and the gorilla with his black skin and hair, his apparent scowl, and his head bulging on top, where the chewing muscles of his big jaw are attached.[1] But cranially and skeletally, the two apes are much alike, their shape differences being dictated to a great extent by the gorilla's larger size. Both apes are very powerful, and arm-wrestling with either is not advised. Chimps are essentially fruit-eaters, while gorillas consume very large amounts of coarse vegetation like vines, soft bark, and wild celery. There are almost cultural differences between different gorilla bands: within the same forest, some prefer some kinds of food, some others.

The orang utan of Indonesia is more of an epicure, going from one kind of fruit tree to another as their fruits ripen in turn. The male orang is about as large as a gorilla and as strong. But he is more completely a tree-dweller, and he is well adapted to such a life, with his very long arms and his flexible legs with good grasping feet. His long hands are almost hook-like and his thumb is short; in order to hold things, he uses his thumb against the side of his index finger — not the best arrangement. He grasps a piece of fruit more by wrapping his fingers around it than by using his thumb, and when he is on all fours he is practically standing on his wrists and ankles, with hands and feet balled up into fists. He is actually seldom on the ground, although in zoos he may become quite good at walking erect, on locked stiff legs and clenched-up feet. This makes an excellent object lesson in how differently he and we have become adapted. He would have the last laugh if we undertook to follow him up a tree.

In looks, an orang is entirely different from his African cousins. A male, in particular, is rather spoon-faced, meaning that his jaws tilt upward in front, and he lacks the forbidding brow of the other apes. With his small eyes and calm, steadfast gaze out of his big milk-chocolate face, he conveys a sense of wisdom and serenity.

---

1. It is not really proper to speak of "the" chimp or "the" gorilla because there are three or four subspecies of each. Two full species of chimpanzee are recognized, one being the bonobo, or pygmy chimp. The several species or subspecies of chimpanzees differ somewhat in skin color, certain facial features, and other characteristics.

Fig. 14. A Young Orang Walking on his Balled-up Fists

According to tests he is as bright as the other two apes and possibly brighter. A mature orang develops flattish cheek pads at the side and a large double chin almost all around his face, making his eyes seem very small and close together. With this enormous face and a coat of long red hair he is a magnificent sight. Alas, he also is endangered, as his forests are being rapidly cut down for tropical lumber.

The Human Pattern

This brings us to that last hominoid, ourselves. More than two centuries ago, Linnaeus installed order — as human beings see order — in the world of life with his system of formal classification. He divided life forms into progressively smaller categories: classes (for example, the mammals), orders (such as the primates), families, genera, and species. He named us *Homo sapiens*, but he could not find any absolute physical trait to justify separating us from apes as he knew them then. So he fell back on the soul as the one thing we claim and apes do not. Today the large apes are placed in one family, Pongidae,[2] while all men, present or past, are placed in another, Hominidae.

Some anthropologists hold, with reason, that the African apes are so close to us that they should be assigned to the Hominidae also, leaving the orang alone in Pongidae. One argument would be comparison with the scale of anatomical differences in other mammals. For example, within the order Carnivora, only a family difference is recognized between all the cats, large and small, on one hand, and all the dogs, wolves, and foxes on the other; and there is certainly much more anatomical difference between a lion and a wolf than between a chimpanzee and a human. At the other extreme Julian Huxley, a great naturalist, proposed some decades ago that we human beings should go into a kingdom of our own, separate from the animal and vegetable kingdoms, because of the way we have stepped out of the natural world through language, culture, and general mentality.

Both views have merit, but are not entirely helpful. Classifying does not change things, but it does allow us to talk about them more clearly. There is a lot of spirited argument among specialists over classification, which can be confusing to non-specialists because the specialists are using it to stake out views of relationship or descent: what is being classified is often views, not animals. And it can seem too fussy to non-players, to whom an ape is an ape is an ape.

The common and traditional view is to place all the great apes and ourselves in a superfamily, Hominoidea, but to keep separate the two families it contains, Hominidae and Pongidae. (Gibbons make a third family, Hylobatidae.) Based on these names, in the language we need here, all apes and humans, past and present, are called "homin*oid*," while humans, separate from apes, are called "homin*id*." This is important; unfortunately the two words are almost the same, but there are rules behind it all. Because of the rules the words mean the same things to everyone. You would think that "human" would be more natural, but it is a little too tricky, as

---

2.   The family name, as is usual, is taken from the earliest-named genus contained in it, in this case *Pongo*, the orang.

we shall see. The word can take various different shades of meaning, and that is bad.

So we need to know our Ids and Oids. How do you go about telling a hominid from some other kind of a hominoid? What really separates us from present-day apes? Brains, of course: we have a three-to-one advantage over apes, who themselves have decidedly large brains among mammals. But when it comes to finding other distinctions, we today are not much better off than Linnaeus. Anatomy? The arguments are endless, and the differences get less and less significant as we go back through the fossils. Language? Certainly this is a root difference in what we have become. But dead men tell no tales; fossils don't speak. So we have to focus on teeth, and on legs and feet. Even arms, shoulders, and chests are not very informative. In these we ourselves probably have a simpler, more basic pattern than the apes.

Teeth

If you look at the three molars in humans and the present-day great apes, you will find the same typical pattern of five cusps on each lower molar and four on each upper, with the same depressions and grooves. Molar teeth are big in gorillas. But those in orangs and chimpanzees are hard to distinguish from ours. In fact, the perpetrator of the infamous "Piltdown Man" forgery was able to make the back teeth in the jaw of a female orang utan seem indistinguishable from human teeth by giving them a little artificial "wear" with a common file. Thus, he created an "early man" with a big brain and with "human" teeth set in an apelike jaw. This was good enough to fool everyone for forty years.

When we move forward along the tooth row, however, things are different. In apes the canine teeth are large and protruding, interlocking in the two jaws. The upper canine comes down between the lower canine and the first lower premolar, shearing against both of them; this first premolar has something of a cutting edge in front and is quite unlike the second premolar, which is bicuspid like both of ours. (The Piltdown forger knew these things and broke all the forward teeth out of the orang utan's lower jaw to hide what would have been a giveaway.) The ape lower canine fits in front of the upper canine, and is received by a gap in the upper tooth row between the upper canine and the incisor in front of it. Ape incisors are rather wide, especially the uppers, and they protrude forward somewhat.

All this adapts the mouths of apes for opening tough fruit and stripping coarse vegetable matter. By contrast, human incisors are narrower and more upright. People of the past, as well as those of the present who still live by hunting and eating coarse foods, lack the "civilized" overbite. Typically, they bite edge-to-edge and wear their incisors and their canines down to flat surfaces. Also, human canines, unlike those of apes, are about the same height as the other teeth. Still, a little furtive observation of friends will show you that many people do indeed have slightly higher canines, without actually looking like Dracula. And the teeth have the same relative positions as in apes, the lower canine closing in front of the upper. (By way of a little lab assignment, I suggest running your tongue thoughtfully over your teeth to note the truth of the somewhat complex description in this paragraph.) All the above facts about teeth serve us well in looking for ancestors.

The New Lower Half

The most important comparisons between apes and human beings have to be made in the lower half of the body. Here, there are particular differences, but they all come down to one pattern of function, that of erect posture in walking and standing — the essential human character.

Look at the pelvis of a mammal quadruped — a dog or a deer. The hip bone, lying parallel to the spine, is narrow and long, with the hip joint in the middle. This gives excellent purchase for the muscles that pull the leg forward and back in running. In the apes, who climb and hang so much and do not run on all fours like general-purpose mammals, the blades of the pelvis are widened and so give muscle purchase for wider, side-to-side movements of the leg. In ourselves the pelvis is not only widened, it is also shortened and splayed out toward the front, like a basin (which is what the French call the pelvis). The spine-bearing part of the human pelvis is pushed back and dropped down, bringing the spine, and thus the center of gravity of the whole upper body, more nearly vertically over the legs.

This is all a rather gratifying and fundamental piece of re-engineering. We do not now know the circumstances of its coming about, but it was probably the first main change from the immediate prehominid ancestor.

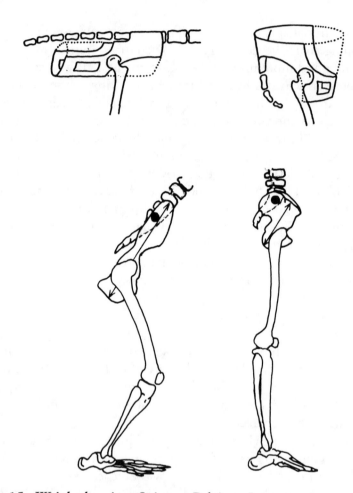

Fig. 15. Weight-bearing: Spine to Pelvis to Leg

Above, an exaggeration of the rotation of the pelvis, from E. A. Hooton. In a quadruped, the pelvis is a horizontal tube, parallel with the spine, with the length giving favorable leverage to muscles pulling the leg forward and backward. The human pelvis has become a bucket, with the spine shifted back and the muscle arrangement less simple.

Below, actual positions of important points. In a quadruped, as above, the axis of the pelvis is roughly parallel with the spine, with the point of attachment forward of the hip joint. In a gorilla, the rear of the pelvis is short but the upper pelvis is long, and the weight-bearing attachment to the spine (*dot*) is well forward, which is disadvantageous for balance in an upright position. In humans, the original axis of the pelvic structure is still inclined, but 1) the spine is bent back, and 2) the part bearing the weight (attachment to spine) has been bent back, bringing it over the hip joint and also close to the joint, making equilibrium far easier.

In making the trunk vertical instead of horizontal, the pelvis did not rotate a full ninety degrees from its original position, and so the spine is pointed somewhat forward as it leaves the pelvis. The rest of the change is made in the lumbar, rib-free part of the lower spine, which curves back to position the rest of the spine and torso vertically above the joint with the pelvis (the sacrum). The shoulders and arms, set out to the sides, and the flat, wide chest also help make the center of gravity more manageable. In us, the lumbar part of the spine has to serve not as an arch but as a column, and a bent one, to support the whole upper body and anything it may be wearing or carrying. If we had been designed from scratch, a better arrangement for our kind of straight-legged biped could surely have been found, or at least one less prone to slipped discs and low back pain. But that is not how things came about.

Our legs have become much elongated in both the upper and lower parts. They have also become more knock-kneed, with the lower legs close together so that they swing back and forth near the center of gravity. This lessens the tendency to sway from side to side as we step on one foot and then the other. When a chimpanzee walks upright (which he does easily but not well), his legs come straight down from hip to ground and cause him to rock from side to side a good deal, and generally, depending on personal style, to wave his arms about for better balance.

More than any other part, the human foot has been altered radically from the flexible hand-like thing of most other primates. Essentially, it has become a third rigid segment of the leg, adding powerfully to the length and strength of our stride. The heel sticks out behind, giving good leverage to the calf muscles behind the ankle joint. The big toe is longer and not thumb-like and is lined up alongside the others, losing its opposability. And the first joint of this toe is brought forward, putting it on a line with the other toes to form the ball, which thus becomes a single joint right across the foot. Add a strong, well-knit arch running all the way from this joint to the heel, and you have a new unbending piece of the lower limb to add to thigh and leg, and thus to the stride. So we have gotten rid of the mobility of the midfoot. (Other mammals, such as horses, have also strengthened their limbs by reducing mobility within the leg, enhancing stability, but have done it in different ways.) When we take a step, we come down on the heel and then lift ourselves the length of the foot arch, shoving off again from the forward joint, the ball of the foot. At this point the toes loyally do a little bit to help.

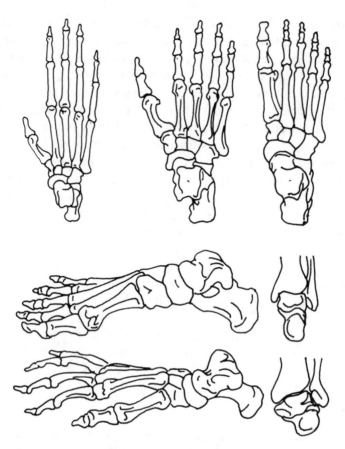

Fig. 16. Human and Ape Foot Skeletons

Above, left to right, orang, gorilla, and human foot skeletons drawn to the same length. In the human foot, the bones of the toes are only a little more than 50 percent of the length, and the point of support of the leg is at almost the 25-percent point, with a long heel. The great toe is powerful and aligned with the others. In the gorilla foot, the great toe is unusually large for a non-human primate and the ankle section is long; all three of the main toe joints are also long. The orang shows the opposite extreme from the human, with a grasping foot. The toes are long, representing more than 75 percent of total foot length; the ankle section is small and the great toe is very short.

Below, profiles of human and gorilla right feet. The human foot (*top*) shows a rigid arch from heel to ball of the foot (at the forward end of the metatarsals, or first joints of the toes). The gorilla foot (*bottom*) lacks the arch as well as the lining up of the first toe joints. In a back view, the human heel (*top*) is vertical and the ankle joint surface is level; in the gorilla (*bottom*) the heel slopes out and the ankle surface in.

Better still, although the arch acts as a new, extra piece of leg, it is not like a single bone. The original separate bones are all there, bound tight with ligaments but still having joints everywhere they meet. The whole thing gives slightly, like a tennis ball when hit or bouncing, and it puts this energy back as the foot lifts off, especially in running. Human feet are admirable things until the arches fall. Then the good of it all breaks down. The tired and the elderly stump around on the flats of their feet, essentially resting on their ankles like other hominoids.

The normal human performance is totally different from that of apes, with their projecting great toe and non-arched foot, a foot in which the weight-bearing point travels gradually forward from heel through midfoot to toes with each step. This must have been the plight of our ancestors, since details of structure show that our foot came from a grasping foot like an ape's or any other primate's. Again, just how the transformation came about, and from what causes, is something we still have to learn. Nevertheless, our feet and legs are the most human things about us and are ultimately answerable for what happened later in our evolution. For one thing, standing erect on our legs freed our hands from any work in moving us about. Satan's mischief is only a little of what idle hands can do.

Asking the Molecules

So here we are. The next question is, which of the living apes are we most closely related to? We know that none of them is a direct ancestor, but we also know that we must have had a last common ancestor with one of them. Which one? This is important, in order to narrow the search for our origins. It has been a quest ever since Darwin.

As past anatomists learned more and more about anatomy, scholars compiled lists of features, big and small, and began to score the several apes for nearness to ourselves. The consensus has favored the chimpanzee, or sometimes the gorilla. Some scholars have persisted in choosing the orang because of a few traits not found in the other two, like a high round forehead without those heavy brow ridges that give the gorilla his menacing look. Actually, totting up separate features in this manner is a rather unsophisticated treatment; instead, we need to determine which traits are the important ones, and what the general pattern of structure and activity may be. For example, an orang has long, red body hair; a gorilla has short, black body hair; and a human male has sparse body hair of different colors, in varying amounts. We do not know the significance of any of this, and so it tells us nothing about relationships. In general structural pattern, however, we are clearly more like the African apes than like the tree-bound orang.

More exacting and informed comparisons, taking many detailed features into account, do indeed make a close trio of chimps, gorillas, and human beings. Particularly important evidence just now is not that of physical traits, but of molecules: DNA (DeoxyriboNucleic Acid, which composes the code of heredity) and proteins, the active substances of living things. This is the result of spectacular discoveries and technical achievements of very recent years. It is a large area and, in fact, the scientific journals are having difficulty publishing all the necessary details in such studies. However, here are the essentials.

DNA (the material of the chromosomes) consists of extremely long paired strands joined by successive pairs of organic bases. Sequences of these are the genes, the pinpoints of hereditary differences, which control the formation of proteins. The genes and their offspring proteins are exactly repeated as cells reproduce, millions of times over unless, due to a very rare error or the action of external forces (such as cancer-causing agents), the wrong base is substituted in a gene. Such a change, or mutation, may have no effect on the protein, or it may have a substantial effect, producing, for example, the abnormal kind of hemoglobin in the red blood cells that causes the sickling trait or, if both genes in the pair are the mutant form, the usually fatal sickle cell anemia.

We know nothing at all about how genes form hands and feet in hominoids. We do know that between these same species, the whole complement of DNA differs by only a small percentage. We also know, by virtue of impressive technologies, how to make comparisons, coarsely or in exact detail, of ape and human DNA sequences or of protein molecules. The point is that, over time, mutations necessarily occur by chance in the DNA. If, on the one hand these have important effects in the body, like the sickling trait, or if they fail in producing normal, necessary proteins, the rate of change is held back by such deleterious effects, because natural selection will operate against the mutant genes. If, on the other hand, the mutations are neutral in effect, they may accumulate at a higher rate, but a rate that is relatively constant, being unaffected by selection. This is the key. We can estimate just about what proportion of a protein's constitution, or of the genes in a part or all of the DNA (whatever we are studying), has by this kind of accidental mutation become different in two species since they emerged from a single parent species. From these results we can draw up provisional family trees, and if there is some kind of gauge of the rate of change, we can estimate the time since the two species parted. That is the so-called "molecular clock."

It was not made in Switzerland and is not guaranteed. There has been and still is a lot of argument as to how regularly it runs. But it has produced some valuable results. From many molecular studies, using various methods or materials, it has become clear that mankind and the two African apes are related to each other more closely than is any of the three to the orang utan. So the three great apes are not a single group set off from us, as was thought a hundred years ago, just because they are all shaped like apes. Then, it was apes versus us; now, it is orangs versus the rest of us. Of course, all four hominoids are separated by only modest molecular differences. But the ground seems to have been cut from under those who have argued for a possible orang ancestor, and thus an Asiatic source, for humanity.

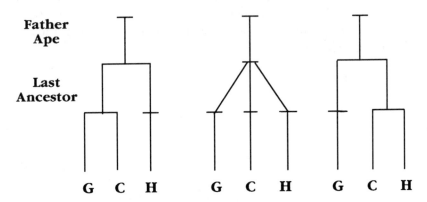

Fig. 17. Views of Ancestral Relations
This diagrams the three most probable patterns of descent for *Homo* and the two African apes. "Father Ape" is the common Miocene fossil ancestor not yet identified, as explained later. This gave rise to intermediate species in the possible patterns shown.

Left, the traditional understanding: gorilla and chimp share one ancestor, *Homo* another. *Center,* the original ancestor gave rise to three species at the same time. *Right,* the present most favored view based on molecular studies: a last common ancestor for *Homo* and chimpanzee, separate from gorilla.

Some molecular tests make all three — gorillas, chimps, and humans — equidistant from one another. A few people have suggested gorillas and humans as the closest pair; but more find chimps and humans closest, with an earlier divide between the gorillas and a chimp-human ancestor. This last view is pulling ahead.

In looking for fossils, it is at least clear which way our gaze should be directed: toward the African apes. Moreover, our separation from them seems surprisingly recent. By comparing rates of change and gene substitutions in the separations among other primate lines (such as monkeys and lemurs), as reckoned from paleontology, the molecular-clock people have figured a separation occurred between ourselves and our last ape cousin possibly five million years ago, or even less. Not long ago, this brought unhappiness to human paleontologists. They had believed, from evidence of earlier fossil apes, that the time necessary for purely human development must have been two or three times that: at least ten or fifteen million years.

None of this is cut in stone just now. There are no clear African fossils to represent immediate parents of the living chimpanzee and gorilla (forests are poor places for fossil preservation). We have found unmistakable human ancestors, whom we shall meet later, but only small pieces of evidence reach back into the murky time zone beyond five million years. It is there that sensational finds of the future will come.

Chapter 6

Primate Ancestors

The age of the earth is about 4.5 billion years. We know this and similarly reliable dates for many events since then, but it was not always so. In the 1600s, Biblical scholarship carefully reviewed the generations since Adam and concluded that the Creation took place in 4,004 B.C.; you can find this satisfying year in almost any Bible, in the notes to Genesis. But that was simply a hypothesis based on the best information around, and it gave way in the nineteenth century when geologists realized that much more time — millions of years — was needed to explain the vast and deep fossil-bearing beds with their evidence of changing, evolving animal life. In the early twentieth century most dates were estimated by counting tree rings or yearly silt layers in glacier-fed lakes (estimates reliable only as far back as a few thousand years) or else by guesswork, most of it quite bad.

New Clocks for Old Times

Why can we now be satisfied that we have reliable time scales? Because of radiometric dating. This started crudely, also early in the twentieth century, but since World War II fresh methods have come along, one after the other, which dovetail or overlap and in general support one another. Typically they take advantage of the radioactive forms of some element. Surely the best known of such isotopes, and almost the first to be used is radiocarbon. Cosmic rays constantly create radiocarbon (^{14}C or carbon 14) in the earth's atmosphere by causing a bombardment of nitrogen 14, but it decays at a regular rate back to its predecessor nitrogen, and so is maintained in a balanced ratio with the amount of inert carbon in the atmosphere.

The rate of decay, or the half-life, of radiocarbon is 5,730 years. This means that, of any given sample, half will decay in 5,730 years, half of what is left in another 5,730 years, and so on until the amount that remains radioactive is difficult to measure. Radioactive and inert carbon are both are taken up, in their standing atmospheric proportions, by all living, growing things. When these die there is a freeze point: the inert carbon remains and the ^{14}C starts to decay. At any given time the ratio between the two can be translated approximately into years since the death of the organism. Ideal materials for such use are burnt wood (charcoal) and charred bone.

The method is extremely useful but only carries back about 40,000 years, until the surviving amount of ^{14}C is too small for accurate measurement. Another important method depends on ^{40}K, radioactive potassium, which decays into ^{40}Ar, the inert gas argon. This decay goes on all the time, but when a rock is heated very hot, any argon in it is driven off and you have a fresh starting point: the new accumulation of argon relative to potassium gives a measure of time. The half-life of ^{40}K is very long, so that the countable range in years does not overlap that of radiocarbon but extends much further into the past. The method is excellent for dating lava and other volcanic rocks, which in many places lie above and below important fossil beds.

There are now other important radiometric methods, but they are complex in requirements and methods, and we will take them for granted at this point.

There is also paleomagnetism. Our compass now points north, but it has done so for only about 780,000 years; before that it pointed south. And before about 2.5 million years ago it pointed north, having oscillated this way back through time. (There were occasional blips of reversal within these periods, lasting only a few thousand years each.) These polarity switches, controlled by little-known events in the earth's core, are recorded in volcanic beds laid down on the earth's surface. They are also recorded in the ocean floor, in lava that has welled up where the earth's tectonic plates (the continental plates) have moved apart. The potassium-argon method is effective in finding the dates for all of these events and has been used successfully to build up a basic framework of dates far into the past.

Thus it is that ages in years can be given to the divisions, major and minor, of geologic time. These periods were

Paleozoic 590,000,000 B.P.
  "Age of Fishes"

Mesozoic 250,000,000 B.P.
  "Age of Reptiles"

Cenozoic 65,000,000 B.P.
  "Age of Mammals"

Divisions of Time

Time since start (millions of years)		Duration
1.7	**Pleistocene** ("almost recent")	1.7
5	**Pliocene** ("majority of recent" forms)	3
23	**Miocene** ("minority of recent" forms)	18
36	**Oligocene** ("few recent" forms)	13
53	**Eocene** (the "dawn of recent" forms)	17
65	**Paleocene** (the "oldest recent")	12

Divisions of Tertiary Time

ordered and named long ago, using the fossil animals they contained as the controlling guide. But now one can say, for example, that the Paleozoic era, the "Age of Fishes," began with the Cambrian period about 590 million years B.P. (Before the Present), when early vertebrates appeared. The Paleozoic was followed by the Mesozoic era, the "Age of Reptiles," beginning at about 250 million years B.P., and by the Cenozoic or Tertiary, the "Age of Mammals," at sixty-five million years B.P.

In tackling the rest of our history we need a finer time frame for the Tertiary, the era which began with the first primates and extends to ourselves. The box above shows its divisions, which originally took their Greek-based names from the proportions of present mammal groups then existing (for example, *paleo* meaning "old," cene meaning "recent," recent referring to groups represented today). This has been constructed by generations of paleontologists and by a lot of recent work in dating. The fossils themselves are not particularly obliging. Only exceptionally can they be dated directly by radiometric methods; their age must usually be estimated by association with datable materials or by the general complex of fossil species in which they are found. Great regions and time zones are still void of information.

An important point implicit in all of the above is that radiometric dating is taking over, more and more, from the time-honored dependence on fossil animal sequences in timing events.

Fossilization

Becoming a fossil is not easy. An animal dies or is killed on the African savanna. At once its flesh is eaten by the killer and by vultures and hyenas. Insects and worms get their share. Hyenas consume most of the bones, perhaps carrying off the skull for later attention. Any bones left over eventually dry and crumble under the sun. In the forest, other things deal with a carcass: wild pigs have been noted scavenging and scattering the remains of both gorillas and orang utans. And on, or in, the acid soil of the forest the bones disappear before long. Even human burials in a climate like New England's disintegrate within a century or so, or where the soil is dry or well-drained, within a few millennia. This is usually not long enough for fossilization, that is to say, for the mineral replacement of the organic fraction of the bone.

For a promising start, bones should land in water and become silted over, followed by the process of mineralization. A skeleton, large or small, may lie somewhere in the flood plain of a river or lake. After flooding by rains, it may be washed into a stream. Here the skeleton will normally be dispersed, some bones traveling with the water better than others: a primate or human skull, being roundish and hollow, can go rolling far away while leaving the denser, hook-like lower jaw behind. It is unusual for several parts of the same skeleton to be found together.

Subsequently, such fossils are normally well covered by later layers of silt, often quite thick, so that hopeful digging from the surface is out of the question. Nature has to expose them, usually by lifting the beds up to a height where erosion will do the work. That is why the Bad Lands in the American West are good lands for paleontologists. Olduvai Gorge in Tanzania, the scene of important fossil discoveries, was formerly a large lake basin where bones became entombed along the shores as the lake silted up. Wind-blown sands also added to the overburden. Then, as in the Grand Canyon of the Colorado River, water action cut the gorge through the many layers of sediment, which go back nearly two million years. Today rains constantly wash the fossils out of the exposed banks. One can only imagine how many wonderful pieces were washed away in the millennia before the Leakeys began looking for them and how many are still buried in the vast deposits remaining all around.

So geography and climate determine whether animals of a given countryside at a given time will get fossilized at all and whether they will ever be exposed to view: they may lie under immense over-burdens of soil or, if not, they may get washed away again. We are very far from having samples of every important period from every continent. In fact, we are unlikely ever to have anything like a perfect record.

Tertiary Primates

The fossil record of primates is comparatively good if far from complete. We are much at the mercy of the fossil lottery. As we shall see, some time ago the abundance of early fossils from North America and Europe created a picture that was partly misleading because of the total absence of fossils from the same period in Africa — which of course did not mean that primates were absent there. The meager African remains now coming to light may upset many previous attempts to read the record.

We know there is a family tree. Imagine a tree inside a house, like a Christmas tree except that it branches out more above than below (see Figure 18). You stand outside; a few windows of the house are open, but they show only pieces of the tree. More windows open, but much of the wall still has no windows, leaving you to deduce how some of the branches connect. In books, we see evolutionary family trees in plenty, but the authors usually forget to show how much of the tree is blocked out by the façade of the house.

As to the primates, today's living species suggest a series of stages leading up from simpler forms (lemurs, etc.) to ourselves. But they are actually end products, survivors from more numerous ancestors, and so they are not necessarily an accurate reconstruction of the past. Nevertheless, primates are served better than most mammal orders. Take horses: their single living genus, one-toed Equus, is recent, and does not begin to suggest the great tree of horses, now extinct, that is known to paleontologists. Over the same time, by contrast, the divisions of the Tertiary era give us primate fossils showing advancing steps that are still represented, at least roughly, in living species. The correspondence is what we will review herewith.

Primates probably became a distinct order of mammals about seventy million years ago, in the late Mesozoic. Then or earlier, there existed a superorder, a bush of first lines, that led to primates, and to ancestral bats, and various insectivores, among others, including some peculiar near-primates, the plesiadapiforms.

Paleocene

These plesiadapiforms prevailed in the early Tertiary. The windows are Europe and North America, continents which at that time were joined together. The many plesiadapiform species can be called "primates" only with quotation marks; they looked like primates, they had many skeletal traits of primates, and they lived like primates, but they are off the line of primate ancestry because of various non-primate features. They had rodent-like rather than primate-like incisor teeth (which nonetheless did not grow continually like those of rodents); they lacked the primates' completed bony ring of the eye socket; and at least some of them had claws. No later descendants are known. The main group is called plesiadapids because they seemed to be "almost adapids," a group that succeeded them in the Eocene.

Eocene

Here the scene is the same — Europe and North America — but the denizens are true primates, also very numerous. One branch (the adapids) foreshadows the lemurs and lorises of today. The other branch, the omomyids, foreshadows the single living genus *Tarsius*, that enigmatic animal who, with a hairy upper lip like us and some other differences, has usually been assigned a very low place among the higher primates. But this assignment has always been controversial, and *Tarsius* remains peculiar and specialized.

The copious Eocene fossils of the north have given the anthropologists much work and little satisfaction. Evidently, the geographical accidents of discovery are at work; it is unlikely that the ancestors of these primates were northern. A few new fossil jaws and teeth from North Africa and China, as well as the presence of lemurs in Madagascar, all suggest that these prosimians actually evolved in Africa and, in a warming climate, migrated to the northern hemisphere. Much is coming to light at the moment.

Primatologists seeking the ancestry of higher primates have felt obliged in the past to choose between the two branches. But no connecting fossils have been found, and in North America some known descendants lasted a long time as progressive prosimians, without signs of transformation to simian status. With much more information today, it is evident that none of the above is ancestral, and that our ancestors arose from still another branch of prosimians.

Oligocene

There is a disjunction in the Oligocene where the windows dim in North America and open on Egypt and South America. In the latter, definite ancestors of the platyrrhine monkeys appear, presenting us with a mystery. The platyrrhines, though clearly separated from the Old World catarrhines, are, and were, perfectly good higher

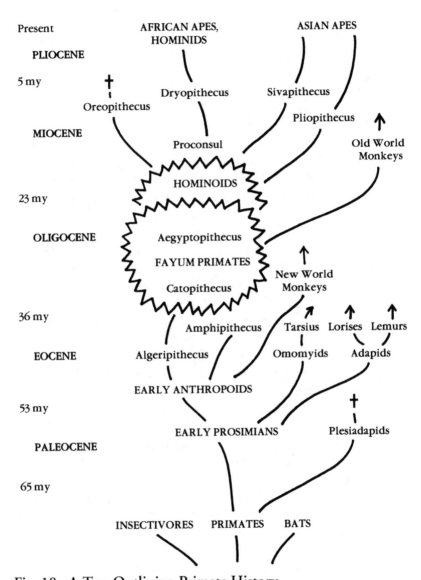

Fig. 18. A Tree Outlining Primate History

This is a skeleton only, and does not attempt to carry all lines to the present, for example New World monkeys. The time scale at the left is meant only to give the beginning ages of the epochs. The diagram emphasizes the active evolutionary radiation seen in the Fayum fossils, and the succeeding multiplication of apes during the Miocene.

primate relatives; they are different in face and teeth, but molecular evidence supports the relationship. How did they get to South America?

An African origin for these monkeys has to be assumed (descent from North American prosimians is out of the question), but the South American and African continental plates began parting so long ago that a crossing of the Atlantic, even by accidental rafting, had seemed impossible. Now, however, the dating of Old World higher primates is being pushed so far back that we can consider the possibility of an earlier split between platyrrhines and catarrhines in Africa and thus of a practical crossing of ancient platyrrhines to a nearby South America. Thus we are reminded again of how much we do not know; luckily, we can go on with the human story without having to settle this matter of the New World monkeys.

Egypt returns us to Africa. We are reminded that this continues as the evident scene of essential primate developments, and probably of origins. We are also reminded that our knowledge of the past depends on the chances of preservation and discovery by the great outpouring of primate fossils of the late Eocene and Oligocene at a single site. Africa is otherwise poor in such sites.

The scene is the Fayum depression, southwest of Cairo. This basin has a lake, surrounded now by howling desert; but in the early Tertiary it was a piece of a coastal plain near the Mediterranean Sea. It was a damp tropical place with rich forest vegetation fed by streams that entombed many fossils. Deep deposits were laid down; those now exposed run from the middle of the Oligocene, about thirty million B.P., down into the Eocene at perhaps forty million B.P.

The Fayum sites are rich, yielding many species of early primates. Some of these apparently went nowhere, or at least cannot be connected with living forms, but some were surely ancestors of monkeys and hominoids.[1] A few give interesting signs of connection with African apes, especially *Aegyptopithecus*. This was the largest of the lot, about the size of a South American howling monkey, which

---

1. To quote Elwyn Simons, the authority on all this, the sites have yielded "cercamoniine adapoids, omomyids, tarsiids, plesiopithecoids, lorisoids, parapithecids, propliopithecids and at least one undescribed primate family." Never mind translation: this arcane language expresses, in whole families and superfamilies, the wealth and variety of these early forms. Simons goes on to say "nowhere else on earth in one small area is there such a diversity of fossil primates."

is not small. In skull and teeth it shows many likenesses to the Miocene apes coming next, namely *Afropithecus* and *Proconsul*. But this would-be ape had a tail, and shows no signs of being a leaper or a swinger. Much earlier was *Catopithecus* of the Eocene, the size of a small monkey. The latter, like most of the Fayum primates, had a tooth count of 2–1–2–3, the same as ours, together with the basic skull features of later apes and Old World monkeys, although the animal itself was much more primitive. Still, *Catopithecus* might be the earliest known primate ancestor of human beings.

Important new find sites have been appearing in Africa and Asia, though yielding only scraps so far. From the Algerian Sahara have come three minute teeth that are recognized as belonging to a higher primate. This wee animal is named *Algeripithecus*. The teeth bear a close resemblance to those of *Aegyptopithecus*, but they cannot tell us much more — whether, for example, the premolar count was 2 or 3. What is important is the date of the animal: it appears to have lived in the middle or early Eocene, perhaps up to 50 million years ago. This gives us a very early anthropoid that might serve as a human ancestor or something near it. But it is in the Fayum, the one known spot with a largesse of Oligocene fossils, that we see higher primates diversifying and evolving, a trend that must have been widespread at the time, at least in Africa. With such a picture now in hand, new finding places giving only modest pieces can carry a great freight of information.

Miocene

Widely in the Old World, the middle Miocene was a time of monkeys and particularly of apes. Unfortunately, there is a gap of about eight million years in our knowledge of the late Oligocene and early Miocene — unfortunate because some important evolution was taking place.

Old World monkeys, with their specialized cross-crested teeth and their efficient quadrupedal running and climbing, have not been clearly traced to any of the Fayum primates. They started their later career slowly, arriving in Europe from Africa about ten million years ago, and reaching India much later, and the Far East later still. They expanded into many species only in the Pliocene, after five million B.P., getting as far north as England and Japan; but with colder weather they vanished from Europe except at the Rock of Gibraltar.

Apes — recognizable hominoids — broadened their territory earlier. Small apes go back to twenty million years B.P. in East Africa. The earliest, known only from an upper jaw from Kenya, is at least twenty-four million years old; thus true hominoids had evolved before the end of the Oligocene. The next of them were not at all

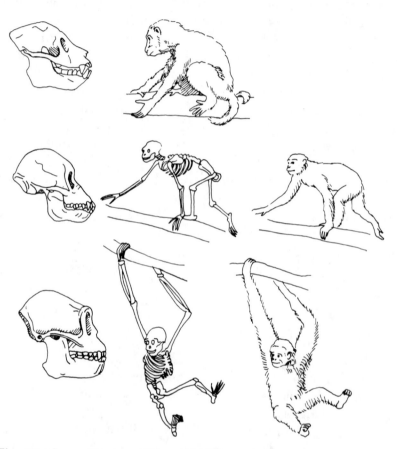

Fig. 19. Some Tertiary "Hominoids"
These three pictured are fairly well known in the skeleton but are not otherwise typical among Hominoidea. *Top, Aegyptopithecus.* This early species was small and had a tail. *Middle, Proconsul.* Tailless and typically hominoid as to teeth, this was an arboreal quadruped, like monkeys, with none of the skeletal adaptations of later apes. *Bottom, Oreopithecus.* Apparently an orang-like brachiator, it was puzzlingly aberrant in tooth form, leaving its affiliation in question.

Fig. 20. The Dryopithecus Pattern

This shows a lower right molar tooth with the classic arrangement of five cusps; fissures on the surface are marked by a Y-shaped main fissure embracing the central cusp on the cheek side. Hominoid lower molars are variants of this pattern; some individual ape and human molars may have six cusps; human molars, in the smaller tooth row, tend to degenerate to fewer cusps.

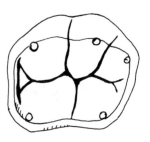

like modern apes but were simple, rather monkey-like quadrupeds. They are well represented in East Africa by *Proconsul*, with an apelike skull and teeth but a non-apelike body.[2] Europe had the slightly gibbon-like *Pliopithecus*.

Later in the Miocene, from about sixteen million B.P. on, we are treated to a fanfare of recognizable apes, recovered at sites from South Africa to Europe and China. The first to be discovered was *Dryopithecus*, found in France in 1856. He had a pattern of cusps and fissures on his upper and lower molar teeth, known to anthropologists and suffering undergraduates as the "dryopithecus pattern." This pattern is faithfully reproduced in today's great apes, in ourselves, and in various extinct apes who are thus clearly shown to be cousins of ours.

Something like two dozen species of Miocene apes have been distinguished. Some scholars would prefer to consolidate, placing them broadly in two main groups. One group, allied with *Dryopithecus*, is primarily European and African and suggests the ancestry of the chimpanzee and gorilla. The other, allied with *Sivapithecus* of Pakistan, suggests ancestry of the orang. But in fact no such relationships are clearly established.

And this simplification is probably too pat. These are mostly "dental apes," known, even well known, only from jaws and teeth,

---

2. Linnaean names are usually latinized forms based on Latin or Greek roots, but are not required to be. *Proconsul* looks Latin, but the fossil was actually named after a popular chimpanzee in the Manchester zoo. Take the case of one of those American fossil prosimians, found in South Dakota. Its finder, an otherwise sane paleontologist, named it *Ekgmowechashala*, which is Sioux Indian for "little cat man."

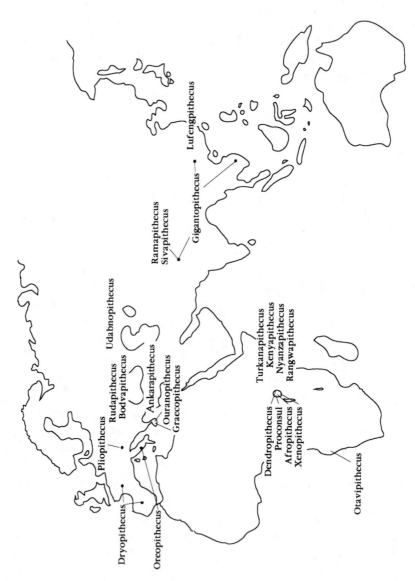

Fig. 21. Where Fossil Apes Have Been Found

This map shows how widely late Tertiary apes were distributed, and how numerous they were. The genera shown do not include all that have been named; and some writers would not consider all shown here to be valid, or to be separate from others. On the other hand, some genera include two or more species, which would increase the actual number of known forms.

the parts of the animal that survive best as fossils. When better parts, such as faces or limb bones, do come to light, they are apt to confound expectations based on the living hominoids (including ourselves). Consider the case of *Ramapithecus*, whose upper jaw was found in 1934. Like another one found later in Africa, this jaw was short and had teeth especially close in pattern to our own. Accordingly, thirty years ago *Ramapithecus* was considered a probable direct human ancestor. But discovery of better parts of the face showed that the species was more likely to be a junior-sized *Sivapithecus*, whose teeth and spoon-shaped profile argue affiliation with the orang. However, upper arm bones of *Sivapithecus* found recently in Pakistan are not at all like an orang's, and leave the question of the animal's gait up in the air (figuratively). We are so familiar with the living apes that we tend to assume that they exhaust the possibilities of ape gaits, which they surely do not.

In fact, the best signs are that Miocene apes were not specialized hangers-from-branches or knuckle-walkers like our living cousins, but instead were more apt to be quadrupedal in trees and on the ground. How, really, were they making a living? We know very little. This raises the question of the decline of apes after the middle Miocene at the same time that the monkeys were expanding.

The Ape Collapse

The hominoids have always been rather generalized, in the sense of tending to preserve an original general form and taking adaptive advantage of its flexibility. This contrasts with the specializations developed in most mammal orders (bat wings, elephant teeth, horse legs, and so on). Apes seem to have been successful both as ground-goers and as tree-climbers, probably surviving on a broad diet of fruits and vegetables.

How did they fall on hard times? Were the Miocene apes actually too generalized for their own good? Did more specialized, agile, quadrupedal monkeys take over a large part of the ape niche? Did gibbons and orangs survive by becoming as strongly specialized in tree life as we now find them to be? Peeking ahead, did we ourselves survive because of a specialized gait? We can only hypothesize. The secrets are still held in the bosom of the family. But at least we see the family.

A couple of other extinct hominoids may have some bearing on the question of specialization and, anyhow, should not be passed over.

One is the highly peculiar *Oreopithecus*. He apparently originated in Africa (where *Nyanzapithecus* is believed to have been a relative) but is best known from Miocene coal beds in Italy. His skeleton, as shown by good specimens, was something like that of a smallish orang: thick trunk, broad pelvis, no tail, long arms with ape-like elbows, and even some human-like feet. So it is written all over him that he was adapted, like present-day apes, for swinging through the branches. In fact, given the dates of *Oreopithecus*, it is more likely that orangs imitated him than the other way around.

But there is a peculiarity. *Oreopithecus* had such distinctive teeth — upright, collared incisors, and molars with a unique pattern — that we cannot possibly classify him with the other apes (hominoid tooth patterns are extremely conservative and similar throughout), and anthropologists now usually place him in a family of his own (Oreopithecidae). So clearly distinct is he, in fact, that some put that family with the apes and some put it with the monkeys. He must have had a long separate history of origin; but he does seem to show, by his converging on them, that the body form of living apes was in the past an advantageous adaptation, since it seems to have been developed independently by gibbons, orangs, chimpanzees, and *Oreopithecus*. Perhaps, in any competition with monkeys, it became an adaptive escape route for them. But *Oreopithecus* and most of the other Miocene apes all became extinct.

The other fossil oddity is the largest primate who ever lived. He was found in 1935 in a Hong Kong drug store by Ralph von Koenigswald, one of the stars in the story of fossil finders. Choice ground-up fossils are used in traditional Chinese medicine, and von Koenigswald made a practice of going through apothecary collections in the Far East. This time he found three very large, obviously hominoid teeth, and he named their owner *Gigantopithecus*. Since then more specimens have been found in China, Vietnam, and India: many teeth and some massive lower jaws, but no other parts. The jaws and teeth have to be seen to be believed: they dwarf those of the gorilla. The teeth, including the canines, are worn down flat, a human characteristic. In fact, it was earlier proposed that *Gigantopithecus* might be a human ancestor, but he is now placed with the Asian group that includes orang ancestors.

The Indian finds of *Gigantopithecus* are about eight million years old, but those from southern Chinese caves go back only a million years or less. In fact, one of these caves has also yielded three human molars, so some early people may have seen *Gigantopithecus* and vice versa. Let us not speculate on their interactions. The caves are in panda country, and microscopic study of teeth shows that, like pandas, *Gigantopithecus* depended heavily on bamboo for food; with such a chewing habit, his large, ground-down teeth make sense. He was clearly not carnivorous, and doubtless was as harmless as today's apes.

It has been two generations since anything as peculiar as *Oreopithecus* or *Gigantopithecus* has come to light. However, new and broader finds of hominoids and other primates, finds more central to the problem of human ancestry, are occurring regularly, to the entrancement of specialists. The problems multiply, and so do the anthropologists.

Chapter 7

The First Hominids

The bygone apes of the Miocene may be telling us that we know less about our origins than we think we do. For two centuries we have been fixated on chimpanzee, gorilla, and orang, and lately on the first two. There has been speculation, for example, as to how we might have converted the knuckle-walking gait of the two apes to our own form, and how our teeth might have followed from theirs.

The choice of models may not be so limited. Those apes are our nearest *living* relatives, but others, before the collapse of the later Miocene, might have been closer still. Consider *Ouranopithecus* of Greece and *Ankarapithecus* of Turkey, both about ten million years old. Good faces and jaws have been found for both of these, and some skeleton bones for *Ankarapithecus*, still under study. Their teeth have been found to approach those of earliest hominids, to whom we turn next. The faces of these apes, although entirely ape-like, are interesting. On the whole they can be distinguished from contemporary orang ancestors but also are not quite as "ape-like" as the African apes. Some dental features of *Ouranopithecus* point toward ourselves: slightly reduced canines and more upright incisor teeth; aspects of the face in *Ankarapithecus* point toward all of the living apes and specifically to none of them; and so on. What the two had for arms and legs, and thus for locomotion, we cannot just now say: were they something like chimpanzees or not? But for skull and teeth it is hard to imagine something more suitable for the ancestor of all.

After these important ancestors, no further direct links to the living African apes are known over the last ten million years. On the human side, however, the gap is about five million years long, after which several kinds of clear hominids begin to appear, showing that there was a bustle of human evolution, apparently in several different lines. Essential changes from apes were on the way: walking on two legs and, much later, far bigger brains. We have extraordinary

Fig. 22. The Laetoli Tracks
Left, part of the trail of footprints of three individuals. On the right, a second individual stepped in the prints of the first; those on the left were made by a child or possibly a female. *Right*, closeup of a few prints, showing the arch and the big toe lined up with the others in a human pattern. Altogether, the prints show a normal bipedal gait, with the feet close together, and with well-marked heel, arch, and toe impressions.

evidence of the first: at Laetoli in Tanzania, Mary Leakey uncovered a remarkable set of footprints having a fairly secure date of 3.7 million years. Two individuals walked side by side, while a third stepped in the prints of one of them. They were walking through freshly fallen volcanic ash, which soon hardened, and it may have been hot; but the single-file habit is characteristic of bare-footed hunting peoples living today. The preservation of the prints is so good as to make possible the most detailed photographic contouring of the impressions. This shows the decidedly human form of the foot and of the stride: a well-formed heel and ball; weight-bearing on the outside of the foot, with the arch on the inside; the big toe in normal position doing most of the work — altogether like prints made by habitually bare-footed moderns. However, there are signs of some crosswise flexion in the arch — it would be incredible that no final improvements should have been achieved since then. Very clearly, we see upright striding, nothing like the four-limbed shuffling and scampering of chimpanzees.

Becoming Erect

This is a first great episode, the crucial switch in our evolution, coming before bigger brains. How did it happen? Chimpanzees and gorillas walk and run erect easily enough when they want to. In the forest, gorillas do this occasionally, but chimps almost never, except on wet ground. However, chimpanzees living in more open country practice it frequently. What was the need to make this the preferred gait, and so to provide the evolutionary push to fix it in the skeleton? Let us remember not to assume that the ancestor started from the chimpanzee we know today.

In the Upper Miocene, about 10 million years ago, East Africa cracked open along what is now the Rift Valley system, with its lakes and rivers. The volcanic activity and land uplift changed weather patterns. East of the Rift, the climate became variable; this caused

Fig. 23. A Chimpanzee Walking Upright
He is managing a good load of bananas.

the forest home of the apes to be replaced by shifting woodland and savannas. So environments and animal life on the two sides of the Rift were separated.

Such an environmental discontinuity is just the situation for an evolutionary discontinuity. This does not completely explain bipedalism, and in fact some of the newly found ancestor forms seem to have lived in woodlands. But note that savanna chimps, as noted above, are more apt to walk erect. With woodland being broken up by grasslands in East Africa, it has been suggested that bipedal running to get from glade to glade might have been a start, like the proposal that lobe-finned fish were not bent on becoming land-goers, but only on moving from pond to pond.

My colleagues, Peter Rodman and Henry McHenry have put this light on it: our bipedalism may be a recovery from the *inefficiency* of the gait of our ape ancestors. The living apes are not true quadrupeds like the monkeys: apes are four-handed, not four-footed. This is fine for forests: gorilla country is tough going for human walking, and passing through a bamboo thicket is something like horizontal climbing. However, those two anthropologists have calculated that, on open ground, our two-legged walking is decidedly less expensive of energy than a chimpanzee's quadrupedal walking. So a new situation, getting across grassland to a new patch of forest, would be better met by walking upright.

Various other behavioral causes fostering bipedalism have been put forward. One is to facilitate carrying of objects, like food and infants; evidently this is useful, as can easily be seen in Jane Goodall's movies of a chimpanzee trying to make off with as many bananas as possible. Another is to have a view over undergrowth, for any reasons at all, such as safety. Still another: an erect body absorbs less heat from the sun. There may well have been important interactions with social life. In any case, the changeover was probably gradual, not drastic, with the ability to use trees declining gradually until walking became obligatory, as it is in ourselves.

E. A. Hooton used to refer to *Homo sapiens* as a "tottering biped." Actually, bipedal running and walking is a quite wonderful gait. It is surprising that other mammals never developed it; many dinosaurs had it and birds got it from some such ancestor. Our two-legged walking is fairly conserving of energy and so allows the covering of long distances, if not at high speed. And our running is effective, at either short or very long distances. It is true that most savanna mammals can run a man down, but the danger has probably been overdone, and in his special persistent way a man can turn the tables, and run down other mammals.

Naturally, this is a story of reasoned speculation, given the lack of direct evidence. Be that as it may, perhaps by a combination of advantages, we can assume that bipedal walking more and more became the gait of choice, with body following behavior in due course. And so the hominids would have become established in East Africa, east of the Rift Valley, just where we find them, in a region now without any apes.

I mentioned, well back, the role of happenstance in evolution. Here, of course, I do not mean a giant calamity such as felled the great reptiles, but only a simple displacement out of a groove. Had our ancestors gone on in forests, we may suppose that nothing would have happened; they would not have turned into bipeds and we would not be here. It did not have to happen. But it did, with enormous consequences.

Meeting Australopithecus

The emerging hominids, known as australopithecines, have left plentiful fossils, from almost 5 million years ago. The finds range from Ethiopia down to South Africa and west to Chad, and many sites have provided parts of a number of individuals. The australopithecines come in two sizes, so to speak: "robusts," so called because of their deep and powerful jaws, and "graciles," so called because they are not robust but are more like what might be looked for in a near-chimpanzee.

That is the general picture. There are variations and changes within both groups, which is where the arguments start. There is so much evidence, and so much missing, that there are good grounds for drawing many different "family trees." Each expert clutches his own, and politely points out the deficiencies in those of others. This does not really becloud the whole tableau of what we know, which is a lot.

Remains show that here, about 4 million years ago, work on the new limbs, spine, and pelvis was largely complete; but these parts, although positively reflecting bipedalism, were far from modern. Teeth were clearly hominid, but with primitive traits. Let us say: if *Ankarapithecus* was an ape with a little of the human about him, then *Australopithecus* was a hominid with a little of the ape

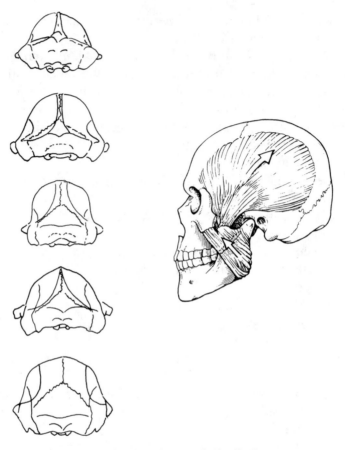

Fig. 24. Brain Size, Jaw Muscles, and Skull Shape

Left, a series of ape and early hominid skulls seen from the back. *Top to bottom*, skulls represented are chimpanzee, *Australopithecus afarensis*, *A. africanus*, *A. boisei*, *Homo habilis*. Starting with the chimpanzee, the breadth of the base becomes relatively smaller and the size of the brain vault relatively larger until, in the earliest *Homo*, it is well rounded. The attachment area of the temporal muscles for the jaw comes close to, or reaches, the midline on top of the skull until, in *Homo*, the larger brain vault and the reduced muscle separate the areas on the two sides.

Right, the two muscles that close the jaw. The temporal muscle (shown here under removed cheekbone) arises from a flat area on the side of the skull and is attached to the forward process on the branch of the jaw. The masseter muscle overlies this; it arises from the lower edge of the cheekbone and is attached to the lower corner of the jaw. This is a short, powerful muscle that can easily be felt with your fingers if you clench your teeth. The skull diagrams give some indication as to how the cheekbones stand out to the side, suggesting the relative size of both muscles.

about him. Somewhere between ten and five million years ago the bridge had been crossed.

While the erectness becomes clear, most of the fairly plentiful African fossils are not bones of the skeleton but skulls and jaws. To grasp the forward evolution in these, consider their working parts. The skull base, with the joint for the spine, faces downward somewhat more to the rear in an ape and the neck muscles attach to the back of it. The jaws project in front, with their length set by the size of the tooth row and, especially, of the canine teeth. Bite comes from the jaw muscles fanning out on the side of the skull; if they are bulky, as in a male gorilla, they need a crest of bone along the top for attachment. Above the eye sockets a crosswise bar of bone finishes off the face and absorbs the force of the lower jaw biting against the upper.

Now put in a brain about the size of a baseball. It can sit on the base and do little to affect shape. In later evolution, as the brain enlarges, the braincase expands over the neck muscles and the ears, and eventually takes over the frontal bar as well. The jaw muscles retreat down the sides, especially as the jaw and teeth themselves diminish.

These are basics. There is much more to the variation and the progress in the australopithecines over the next three million years, from their somewhat primitive appearance to the establishment of the first real, full-sized people. Here is the story of discovery.

The Finders

A brilliant young Australian, Raymond Dart, finished his medical studies under the English greats, Elliot Smith and Arthur Keith, and took up the new professorship of anatomy in Johannesburg at the age of twenty-nine. The next year, in November 1924, he was given a piece of limestone with a provocative-looking fossil embedded in it. By February he had winkled it out (partly with his wife's knitting needles), finding much of the skull of a child, with a natural cast of the whole brain. He saw it to be ape-like but with human features; it was small-brained but was rather more human than chimpanzee in certain comparable features of the brain. It already had its "six-year" molar teeth, and these were squarish and hominid-looking (although nobody said "hominid" or "hominoid" in those days). And the shape of the back and base of the skull suggested a being who walked erect. Dart promptly published all this in the British journal *Nature*; he gave his view that this creature was truly intermediate between ape and man, and he named his find *Australopithecus africanus* (African ape of the south).

This should have caused the sensation it deserved. But these were the bad old days of the early part of the century when anthropology was not very professional. Men like Elliot Smith and Arthur Keith were anatomists who did anthropology on the side. Worse, they were still in thrall to the Piltdown fake, which "proved" that the first men had ape jaws but large brains instead of the other way around; as Elliot Smith then said, "the brain led the way." Both these men scorned *Australopithecus* as just another ape. So did the leading American, Aleš Hrdlicka at the U.S. National Museum, who had not even fully accepted the Piltdown "man." *Australopithecus* did not come into his own until midcentury, although Dart and another determined stalwart, Dr. Robert Broom, kept looking for, and eventually finding, further fossils.

Raymond Dart contemplating the bust of Robert Broom, who is sculpted contemplating the best skull of *Australopithecus africanus*. The bust, a memorial to Broom, is located in Sterkfontein at approximately the same spot the skull was found.

Broom, a crusty Scot, was an M.D. practicing in South Africa, but also already well known as a paleontologist (mammal-like reptiles were his expertise). He was indignant at the treatment Dart had been given, and said so. He began visiting a lime-processing works at Sterkfontein, near Johannesburg; fossil bones were being found in the fill of old limestone caves of a likely age, and the mine foreman was selling the bones to tourists. Quite soon new pieces of *Australopithecus* began coming to light, for which Broom paid the foreman a pound or two.

In the 1940s, Dart did the same thing at another limeworks cave, in the Makapan valley, with even greater success. This was in spite of the foreman there, a strict fundamentalist. Alerted by the gathering publicity about *Australopithecus*, this man ordered the workers, should they find anything that looked remotely like an ape-man, to put it straight into the lime-kiln and burn it. To no avail: Dart and his students searched the waste dumps, where the miners had thrown cave fill of poorer quality lime but richer in fossils, with great reward. And by this time Sterkfontein was a protected fossil site, with Broom blasting away full time, also with great reward.

Already the dike of foreign indifference had begun to leak. William K. Gregory, a major paleontologist at New York's Natural History Museum who first defined the "dryopithecus pattern," came out in 1938 to have a look at what Dart and Broom then had in hand. He was converted: he coined the term "man-apes" for the finds, and created the sub-family Australopithecinae to accommodate them. But all, like Dart in 1925, still thought of them simply as intermediate. And Broom was calling them "ape-men." Then in the late 1940s Dart and Broom both found fine skulls proving the erect positioning of the head, and Dart recovered the first pelvic parts showing the same thing, and it was all over. By 1950 the sub-family Australopithecinae was placed firmly in the Hominidae; and Keith, in his old age, handsomely declared that in 1925 Dart had been right and he wrong. So that controversy came to an end.

Another Time and Place

Ten years on, a new phase of discovery began, this time in East Africa. Here the fossil deposits are lake and stream beds, not the limestone cave linings of South Africa. And here there are dates, whereas no way has been found of telling absolute time in the caves of the south. In East Africa, the deposits are interleaved here and there with volcanic layers, which can be read by the potassium-argon method.

Olduvai Gorge, in Tanzania, was once a lake basin that gradually silted up; then its layers were cut down again in gullies and deep canyons, and rains now wash fossils out. The bottom of the Gorge is volcanic rock, dated to 1.9 million years ago. Louis and Mary Leakey set up shop at Olduvai, and in 1959 Mary came across a splendid skull of a robust australopithecine, near the bottom and so datable to perhaps 1.8 million years. The next year the Leakeys found important other parts, of a different kind. That was the beginning. The senior Leakeys continued discovering fossils at Olduvai, at Laetoli, and elsewhere; their son, Richard Leakey, and his co-workers have had great success on the shores of Lake Turkana; and American-French teams have prospected northward in the Omo Valley and, brilliantly, in the Afar region of Ethiopia. The collections are opulent in amount and in variety.

Controversy is also opulent, and brisk, and not always exquisitely polite.[1] My effort here is to boil it all down, giving not a picture that all, or most, would agree to, but only a general sense of the situation.

Line One: The Graciles

Earliest — at the moment — at almost 4.5 million years, and deserving to be called a missing link, is *Ardipithecus ramidus,* found in Ethiopia. This is given a whole new genus because of its mixture of chimpanzee and hominid traits in the teeth and fragments of armbones, while there are hints only, in parts of the skull, of its being a biped. The "chimpanzee" aspects are to be looked on as leftovers

---

1. To quote J. S. Jones, writing in *Nature*, "Paleoanthropologists seem to make up for a lack of fossils with an excess of fury, and this must now be the only science in which it is still possible to become famous just by having an opinion." Andrew Hill, in the same journal, has these ironic principles for paleoanthropology: "every new fossil hominid specimen is the most important ever found and solves all known phylogenetic problems; every new hominid specimen is completely different from all previous ones, no matter how similar; every new hominid specimen is a new species, and probably a new genus, and therefore deserves a new name." And David Pilbeam, in lecture: "... there is an inverse relationship between familiarity [with fossil material] and expressed certainty of opinion."

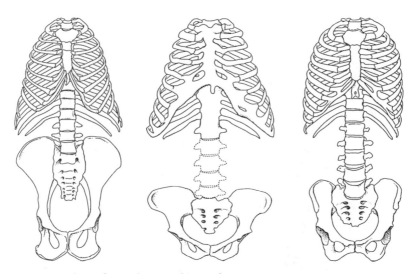

Fig. 25. The Afar Pelvis and Trunk
Broad and flat, the Afar pelvis (*center*) is quite different from that of a human (*right*) and a chimpanzee (*left*); it is, however, adapted for erectness. The thorax is funnel-shaped.

from the ape-human ancestor, while the hominid traits look like an early start in the really human direction. We look forward to finding more of *Ardipithecus.*

Then comes *Australopithecus kanamensis,* barely known from a few parts recently discovered in northern Kenya. It is clearly australopithecine, with a perfectly hominid elbow joint, but hardly anything can be said except that it stands as a new species.

Much more can be said about *Australopithecus afarensis,* familiar as Lucy[2] and family, from the Afar depression in Ethiopia (found from 1973 on by Donald Johanson's team) and from Laetoli in Tanzania (where Mary Leakey had long worked). The age of all these is about the same as the Laetoli footprints, 3.7 million years, and continuing to perhaps 3 million years ago. Much of Lucy's skeleton was found: she was a little over three feet high but other individuals were markedly larger. This may be a sex difference. They were

---

2. Christened from "Lucy in the Sky with Diamonds" (LSD for short). The expedition hammered the evening air with recordings of this Beatles song. According to co-leader Yves Coppens, the Ethiopians preferred to name her "Birkinesh," meaning in effect "you are a personage of consequence."

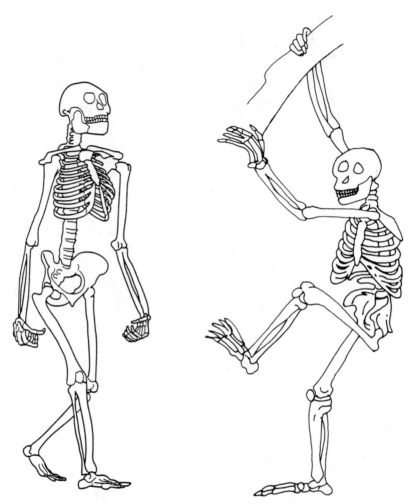

Fig. 26. The Afar Skeleton in Two Postures

Henry McHenry, whose figures these are, has examined in detail bones of all parts of the Afar skeletons in comparison with moderns. His finds:

Primitive traits, over fifty major and minor. Particular ones include a shoulder joint facing more upward, as in climbers; slender, curved fingers; hip bones less curved around toward the front; short thigh bone; and long mid-toes.

Hominid-like traits, a similiar number. These include small arm bones; shortened fingers; a generally human-like pelvis with broadened sacrum and lumbar curve; knee joint with human angles; straight lower leg; well-developed heel and ankle bones; shortened toes; and a big toe lined up with others.

unquestionably erect: their pelves, spines, and knees show it. The pelvis, however, was by no means modern, nor were the feet: the toes were more curved than ours. The heel bone lacked our stabilizing tubercles; and a couple of small ligaments that, in us, tighten the arch from underneath, were apparently not present in these Afar hominids. Finger bones were curved, as in tree-climbing apes. The conclusion is that Lucy would have been smarter in trees than we are today. And there is much agreement that Lucy's gait is not properly understood, and that it was not something simply transitional to ours, although it must have been successful in serving Lucy's purposes. Remember that one of Lucy's purposes was almost certainly sleeping in trees for safety's sake.

Here is something of an enigma. Excellent evidence of a very modern foot, from the footprints at Laetoli. Excellent evidence of feet, hominid but not fully modern, from the Afar bones. Russell Tuttle of the University of Chicago, a leading expert on hominoid gaits and limbs, finds that all aspects of the footprints, especially toe proportions, are remarkably like modern feet, but that the Afar feet are significantly different. He has suggested that the prints were made by a different hominid species, which is certainly a reasonable suggestion given the recent discoveries.

Others have the same doubts that Lucy, or anybody she knew, could have made the prints. And from all evidence the foot skeleton

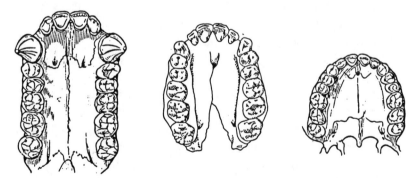

Fig. 27. Upper Jaws in Gorilla, *Australopithecus afarensis*, and *Homo*
This shows the much-diminished canine teeth and the shorter, more rounded arch of the Afar fossils (*center*) compared with those of a gorilla (*left*). The Afar jaw still had a pre-canine gap and the fore part of the jaw projected, unlike *Homo* (*right*).

of *A. afarensis* continued much longer in its unfinished state (to call it that). Nevertheless, Lucy's great toe was lined up with the others, like our own. And ankle and foot bones have some aspects of shape that suggest apes, but also other aspects, studied by exacting geometrical analysis, that could be found only in bipeds. In fact, the signs are that the Afar beings (*A. afarensis*) had reached their own satisfactory mode of walking, one lasting some two million years. The enigma remains in force.

Never mind: the Laetoli/Afar creatures, as known from the fossils, were clearly erect-walking australopithecines. However, their skulls and teeth were also definitely primitive for hominids. The face, in the best skull, was large — broad and projecting, apelike in general form, by any standards. Canine teeth were somewhat projecting, and their premolar companions were not fully bicuspid. There was a small gap in the upper tooth row into which the tip of the lower canine fitted. (But this is not a positive mark of the beast — a similar gap existed in some of the much later Java men, who were far more "human.") Neck muscle attachments were likewise somewhat more primitive, along with some other traits.

Next, over the following half million years, apparently from 3 to 2.5 million B.P., come the many pieces of *Australopithecus africanus*. They comprise finds of Dart and Broom in South Africa; East African specimens are not definitely known. Like the Afar skulls, these have a superficially chimp-like profile, but they are more obviously hominid at first glance. They have a slightly larger brain cavity (about 450 cc versus 400 cc). They have less heavy brows, smaller incisors and canines, and teeth worn flat. An essential difference: the first lower premolar has two cusps like ours, not the near-canine shape of apes, which is a form still hinted at in *A. afarensis*. Another difference: there is no gap before the canine in the upper tooth row. In simplest terms, *A. africanus* looks like a small advance, along the hominid trail, from *A. afarensis*. But — and here is a surprise — the skeleton, as far as known, was definitely more chimpanzee-like than that of the earlier Afar creatures.

Add all this up: four species over two million years plus the extraordinary Laetoli footprints, as well as a lonely lower jaw far to the west in Chad, 3.5 to 3 million years old and currently known only as "Abel." This is no single evolutionary line; instead, it is a general phase in hominid development with varied species taking slightly differing tracks.

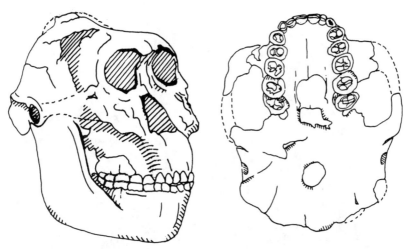

Fig. 28. The Skull and Palate of a Robust Australopithecine
Left, facial view shows the massiveness of the jaws and the existence of a small crest along the top, where part of the jaw muscles were attached. (This is the skull of the Dear Boy, to which has been added a lower jaw as known from other specimens.)

Right, basal view shows the flaring cheek arches (as restored) to accommodate the large masseter and temporal jaw muscles. Especially important is the contrast between the small front teeth, including the canines, and the cheek teeth, emphasizing the specialized nature of the dentition, and probably of the diet, in the robust line of australopithecines.

Line Two: The Robusts

Finally there are the "robusts." Broom found the first ones in the 1930s, two variants at two sites within strolling distance of Sterkfontein, and he named them *Paranthropus robustus* (actual source of the use of "robust") and *Paranthropus crassidens*, in honor of their big teeth. (Broom named genera and species at the drop of a hat; he named his Sterkfontein specimens *Plesianthropus transvaalensis*, but this has been sunk into *A. africanus*.)

Most spectacular was the Leakeys' 1959 discovery at Olduvai of the beautiful skull that started the train of East African discoveries. Because he suddenly fulfilled their long-cherished hopes of finding hominids at Olduvai, the Leakeys called him the "Dear Boy." Other excellent jaws, teeth, and skulls of the same kind have been recovered at Olduvai, Lake Turkana, and elsewhere.

Louis and Mary Leakey contemplating the Dear Boy [the robust *Australopithecus (Paranthropus) boisei*] in 1959 shortly after the pieces were first assembled.

The Dear Boy was promptly christened the "Nut-Cracker Man" in the newspapers. Actually the robusts were virtual milling machines. Microscopic examination of pittings on their tooth surfaces, compared to those of the other australopithecines, shows that they were equipped like *Gigantopithecus* for tough materials, probably seeds and coarse vegetables. To a brain case of australopithecine size (about 520 cc) were attached massive jaws. The back teeth premolars and molars were very large,[3] in striking contrast to the front teeth canines and incisors which were very small; quite the

---

3. With their characteristic openness, the Leakeys brought the Dear Boy, two weeks after finding him, to a congress of prehistorians then meeting in Kinshasa. At a small gathering they put the pieces on a table for handling by a half dozen or so specially interested anthropologists. My wife had a U.S. penny with her, which just fitted a molar crown in size. A penny is not worth much, but for a hominid tooth size it is very big.

opposite from other australopithecines. Tooth surfaces were worn flat. The size of such jaws meant that they protruded somewhat, but they were actually tucked in as far as practical, which meant that they must have had enormous biting force in the back teeth. The cheekbones, anchor for the masseter muscles, were large and wide, leaving the face dished in between them. The temporal muscles, lying underneath the masseters, gave rise to a ridge along the midline of the skull's top, as in a gorilla, but with this ridge placed more forward because the bite was more vertical. The joint sockets for the lower jaw were set out to the side rather than being underneath. The whole is a striking adaptive pattern.

It is the more suggestive that the robustness seems to be all in the chewing apparatus. The animals themselves do not seem to have been giants, and such skeletal parts as are thought to belong to them attest to a body size something rather less than our own size. The skull, apart from the jaws, is not heavy, with a thin-walled vault and with a lot of air cells filling parts that thus look stouter than they are. This seems all the more to stress the special aspect of a dietary adaptation affecting simply the jaws.

Much disagreement has swirled around the robusts. Some see as many as four species; some see two at most (north versus south). In any case they are generally placed in Broom's genus *Paranthropus* following the lead of Bernard Wood at the University of Liverpool. This makes sense considering that the difference from other austra- lopithecines looks at least as great as that between chimp and gorilla. From present evidence the robusts first appear two and a half million years ago,[4] and over the million and a half years of their existence in East Africa they show little change in general nature including brain size; the later ones became more typical, so to speak. So I think the conservative probabilities are that the robust line diverged early, as something that the new hominid potential could produce, that the robusts found their niche, one that did not encourage change, and that they continued contentedly on until human competition or climate shock did them in. They simply show that not all hominids had to become "human."

---

4. This is the date of an excellent specimen, WT 17000, known as the Black Skull. It has the main robust character and size but seems less evolved toward the later examples.

Chapter 8

Enter Homo

A first great happening, erect posture, produced hominids. A second, a critical brain expansion, produced what we can call human beings, whom we place in the genus *Homo*. This grew naturally out of what went before. In any case, by 2.5 million years ago a pair of important things had happened, having to do with brains and tools.

First, evolution was on the march again with what seems like a radiation of new, advanced hominids. Some kind of *Australopithecus* was losing various primitive traits of the skull and, more important, the brain was expanding, from about 450 cc to between 600 and 800 cc.

It is an interesting scientific story. Recognition of early *Homo* was neither a sudden, forehead-slapping insight nor an eye-opening fossil find like the 1959 robust at Olduvai. Rather, it was a gradual taking-shape of perceptions of material, new and old, aided by improved dating of what was already in hand. It began at Olduvai with the recovery, from 1960 on, of various non-robust fragments of skulls, hands, jaws, and feet, followed by numerous specimens from Lake Turkana, all suggesting larger brains but being on the whole rather varied and baffling. (Except for some unreadable fragments, all these species are younger than 1.9 million years.) These parts so resembled the gracile *Australopithecus* that to put them in that genus seemed at first like the most comfortable thing to do. Distinguishing real representatives of *A. africanus* — fossils like the South African specimens — from something more advanced was getting to be difficult, almost like cutting a spectrum.

Homo habilis Tiptoes In

Then, in 1964, Louis Leakey, along with Phillip Tobias (Dart's successor) and John Napier, formally proposed that the degree of apparent change in these Olduvai fossils, small though it seemed, sufficed to place most of the specimens in the genus *Homo* as a new species, *Homo habilis*.[1] Such a lowering of the threshold of admission called for a whole new definition of *Homo*. This seemed like tampering with scripture and was greeted with catcalls. These continued for twenty years, but recovery of similar fossils was continuous. In 1994 a fine upper jaw was found at Hadar (the Lucy area), well dated at close to 2.33 million years. All agree that its short-faced form can only be that of *Homo,* not *Australopithecus.* Thus the appearance of *Homo* at a new and early level is now generally accepted as marking an important divide.

But an inescapable problem persists: the *Homo habilis* collection contains too wide a variation for a single species. Bernard Wood, a primary authority, has examined the material from all sides, including quantitative analysis, and concludes that two species should be recognized: *Homo habilis* and *Homo rudolfensis*, both the same age and both at a primitive *Homo* level.[2] The problem: what evolutionary processes and events were producing these early post-australopithecines?

Finally, in 1992, a new bone fragment joined the collection, a strip from the right side of the skull base running from the mastoid process to a point in front of the joint for the jaw. The piece had actually been found in 1967, in the Baringo region of Kenya. At that time Phillip Tobias noted some features that seemed *Homo*-like but given the state of knowledge then, he was doubtful about positively allocating the specimen. Yale's Andrew Hill and colleagues recently dusted it off again and were able to distinguish it from all australopithecines gracile and robust. They are confident that it belongs in *Homo* and, interestingly, they are inclined to place it in the *H. rudolfensis* branch. But the special importance of the specimen is its date. Careful potassium-argon analysis of the site places the fragment at 2.4 million years B.P., which agrees with the Hadar

---

1. The species name, proposed by Dart, suggests "clever with the hands."

2. The species name, *Homo rudolfensis* (from Lake Rudolph, now Turkana), was suggested earlier for the ER 1470 specimen by the Russian Valeri Alekseev. Note that this is by no means simply capricious naming of the old-fashioned kind but instead a serious attempt to clarify matters by careful description of important distinctions between carefully defined groups of specimens, as well as by care in application of names already suggested.

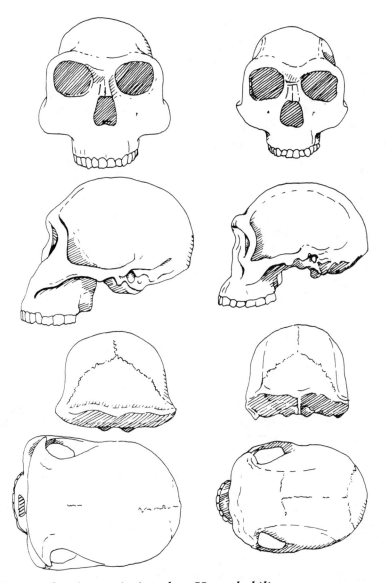

Fig. 29. Specimens Assigned to *Homo habilis*
These two skulls differ noticeably in appearance and in some details such as length of face. However, they are contemporaneous and share other general characteristics, such as enlarged brains, compared to *Australopithecus*. They illustrate the vexing question of whether they represent a single species or two separate species arising at the same time from an australopithecine base. Bernard Wood distinguishes them in some facial features, and assigns ER 1470 (*left*) to *Homo rudolfensis* and ER 1813 (*right*) to *Homo habilis*.

upper jaw in pushing back the known age of *Homo* from a previous limit of less than 2 million years.

Homo at this level remains difficult to define because of the variation among the many assigned specimens, most of them rather fragmentary, recovered from Olduvai, Lake Turkana, and South Africa. Certainly the brain of *Homo* was larger than that of *Australopithecus*, the skull was higher, and the face was more vertical. In at least some cases the cheek teeth were relatively smaller, especially the third molar, although in *H. rudolfensis* the teeth are larger than in *H. habilis* and the face is notably broad. The skull joint for the spine (foramen magnum) was placed further forward than in *Australopithecus*. All of these things are progressive; and complex analysis with measurements seems to show that the forward part of the skull, taken as a whole, clearly should be placed between the earlier and later hominids. However, the skeleton, poorly known at first, now appears to be surprisingly primitive, at least in the case of *Homo habilis*. It is more like the skeleton of *Australopithecus* than like those of later men: the arms are described as long and powerful, and the hands and feet as reflecting earlier forms. This is hardly dismaying; the forerunners of *Homo habilis* had been prosperous walkers for more than a million years.

East Africa is the probable arena of human emergence. It is surely the cockpit of the argument about it. The outline I have given would suggest a simple progression over a period of a million years from the gracile australopithecines to early *Homo*, with the robusts lurking off by themselves somewhere. But the story is not so simple. What we have are still only gobbets of evidence in time and space, not the whole panorama of action. And there is a large number of experts in the field; they know a lot more these days and so there is more to argue about. Their views and methods differ from one to the other. Some scholars discuss from the standpoint of the overall patterns of form and life suggested by given fossils. Some look at precise aspects of musculature suggested by the bones and what they mean. Some take numerous measurements or make meticulous counts of precise features, using advanced mathematical solutions to suggest relationships. Finally, there are natural temperamental differences — these scientists are human, after all. Some want to emphasize likenesses, some prefer differences; and so most of them acknowledge being either lumpers or splitters.

They are not captiously tossing off opinions. Five or six kinds of family tree have been proposed: they have robust species developing out of one or both gracile species, or arising further back; *A. africanus* being descended or not descended from *A. afarensis*; the latter being descended or not from *A. anamensis* (quite *Homo*-like in a few traits), and so on back to *Ardipithecus; Homo* being descended from any or none. It is much more likely that there was a bush of species, not a straight line. Never mind that here; in any case, new finds like the Black Skull or the Baringo fragment torpedo one or more of these schemes. From above the battlefield we can simply take an overview of this whole phase of australopithecine history. Remember, not all the fossils have yet been found.

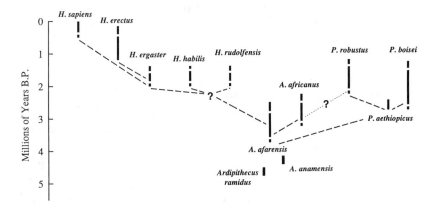

Fig. 30. A Family Tree for Hominids

This follows trees drawn up by Bernard Wood from the latest information and study. Heavy dashed lines indicate the most likely connections at present. Time is shown at the left.

The species is *Australopithecus afarensis*, with possible forerunners *A. amanensis* and *Ardipithecus ramidus*. Two adaptive lines diverge. Very large back teeth and small front teeth mark the heavy grinding adaptation leading to the robust *Paranthropus* species — the early WT 17000 (the "Black Skull"), shown as *P. aethiopicus*, may be the antecedent of both. The other branch, with enlarging brains, but conservative as to teeth, leads to all species of *Homo*. *Australopithecus africanus* appears to have no traceable descendants.

Various alternative schemes are proposed by other workers. R. Skelton and H. McHenry maintain, from a careful compilation of individual traits, that earliest well known *Homo* is closest in descent to the robusts (*Paranthropus* in the tree above).

Tools and Food

The second new thing is the appearance of recognizable though very crude stone tools. The earliest are dated at almost 2.5 million years, and this was a problem when we believed *Homo* to be half a million years younger than that. But now, after restudy of the Baringo fragment, *Homo* and stone tools fit together like hand and glove. That is the primary importance of the Baringo fossil. The recently found Hadar upper jaw, at 2.3 million years, is also associated with Oldowan tools.

Some probable bone tools were also present at this time, showing marks of wear. Anthropologists have mimicked these marks by fashioning similar tools of fresh bone and digging up edible roots and bulbs with them. As to the stone tools, they are no more than pebbles broken so as to produce a sharp edge; given that hominids have no claws or fangs, such tools would have been very useful for cutting into a dead animal's skin or taking off meat.

Microscopic examination of these tools, many of which only a trained eye can recognize as tools, can be very revealing. With stone of a suitable nature, it becomes possible to distinguish among the kinds of smoothing that result from cutting meat, scraping wood, and cutting grass and other vegetables. This is accomplished by modern experiments making the same tools and doing the same things with them.

As to the toolmaker, he can only have been *Homo*. Any gracile *Australopithecus* apparently had already gone into extinction. And the betting is against the still-surviving robust Dear Boy as a toolmaker. He is presumed to have been a vegetarian: he seems quite unequipped for eating meat, with his big flat-worn molars and his relatively tiny front teeth. (Nevertheless, his hands were capable of skillful manipulation, and he probably could have used tools well, whether he made them or not.) The teeth of the gracile *Australopithecus* were closer to the main line, with larger incisors, and *Homo* kept this pattern.

The important thing is meat. Chimpanzees eat meat (occasional monkeys or gazelles), but fruit is their staple. The habilines doubtless depended largely on vegetable stuffs, but it is clear from their tools and food bones that meat was a regular component of their diet. Not that they were mighty hunters. It is probable that they did more scavenging, cutting meat from carcasses of animals killed by other animals. In a few cases, careful examination of food bones from the hominid sites shows marks made by carnivore teeth and marks made by sharp stones (tools), both on the same bone. On some of these the tooth marks are superimposed on the tool marks or vice versa, depending on who got there first; in such cases, both visitors were probably scavenging. Bone studies show signs not of carcasses being cut up, as if the habilines had actually killed the prey, but of meat being cut off the parts left over by a hunting animal. The hominids were not epicures, but they were indeed getting the benefits of meat in their diet.

This likely role for early hominids has been experimentally acted out by G. G. Tunnell of the University of Auckland. He picked a spot in Kenya with no human inhabitants and with few animal species and patrolled around it for months on several visits. He found that contact with lions would have some hazards but plentiful rewards. Lions tended to kill prey fairly frequently, gorging on the viscera and then moving away to rest. Tunnell's selected pride of lions usually left, temporarily unguarded, enough high quality, unspoiled meat to feed a band of ten hominids full time. Simple stone tools could cut hides, which vultures could not do, and even plain rocks could bash open skulls for the brains, which lions could not do. Lucy and her ilk could save their own hides by climbing and sleeping in trees, at which they seem to have been adept. Apparently, making a living by scavenging would have been highly practical for tree-climbing australopithecines and also for hominids living permanently on the ground, after they became brainy enough to deal with predators. This would have been a way of life for the gracile australopithecines but not for the robusts, whose teeth apparently were not those of meat-eaters.

With the appearance of detectable stone tools we see a juncture in history. But culture clearly existed in our anthropoid past. Chimpanzees, as Jane Goodall found some time ago, will make "tools": they will strip leaves from a stalk, for example, and poke it into a hole in a termite nest to fish out the termites that attack it. More recently chimp communities have been studied across Africa, showing surprisingly varied "cultural" patterns. Drinking may involve a number of different plant materials used as a sponge. Hard-shelled nuts are cracked using a suitable stone perhaps carried intentionally to the tree; this is not easy work, and mothers have been seen showing their little ones how to do it.

Certain kinds of fruit enclosing seeds or nuts are eaten widely, but only in four neighboring groups in West Africa do the animals crack the seeds in the fruits for the kernels; Harvard's Richard

Wrangham suggests that perhaps one bright chimp invented this idea and that so far it has spread, by learning, only to these few communities. Some chimpanzee groups seldom hunt; by contrast, one group was seen hunting monkeys with great purposefulness, two or three chimps vociferously driving a monkey group into the arms of other chimps already carefully and silently posted at the likely escape routes. (Of course, some other habitual carnivores use such tactics.) "Culture" is not defined by the use of tools or even by the transmission of ideas. What is cultural rather than "instinctive" is variety in the ways of doing, or not doing, these things.

Australopithecines must have been as inventive and culturally resourceful as chimpanzees. But in the habiline record we see culture more clearly: a patterning in the reshaping of blunt stones into sharp ones, carried out wherever a tool is needed, with many implications as to skilled hands and the teaching of one individual by another. This belongs more to the nature of human culture, continuous though it may be with chimpanzee activity. At any rate, the habiline stone implements mark the beginning of the record of toolmaking and of a species-wide cultural heritage. And it has been persuasively argued that they also mark the beginning of language.

Thinking and Speaking

Australopithecines had been getting along with roughly ape-sized brains. Habilines show the first step in the brain growth that has continued to the present. Erectness was the underpinning that made us hominid, but, of course, it was brains that made us human.

It must be more than a coincidence that tools and early *Homo* appeared at the same time. Brains continued to expand, and tools continued to become more varied and better defined. Is that the whole story? Evolution is parsimonious. At any given point the toolmakers were already successful animals, and we assume that brains would have evolved to be no larger than was necessary for purposes at hand.[3] Why, then, did the increase continue? Surely

---

3. Earlier I mentioned preadaptation as a factor in evolution, as when air sacs in fishes became lungs in land animals. Brain size getting ahead of whatever behavior was making it useful would not be the same kind of thing; instead, it would exemplify orthogenesis, the now-abandoned idea that evolution can get into a groove of momentum all its own.

because there is more to "intelligence" than toolmaking. Other powerful "tools" such as social interaction and communication would, I think, have added survival value to increased brain capacity.

We have to consider the evolution of language. Here, of course, there is no hard evidence; the beginning of speech cannot be pinpointed. But we may assume that some kind of behavior laying the groundwork for language was in being by this time. The survival value of communication is enormous, and language is the key.

We can learn absolutely nothing from living peoples about language evolution. The most "primitive" of these, whether beetle-browed Australian aboriginals or culture-thin Fuegian Indians, speak languages every bit as elegantly inflected as English or Chinese — often more so, and typically used more correctly.

Nor have apes been of much help. Chimpanzees and orangs are simply not equipped to speak like us. In them, the larynx is too close to the mouth to allow a free flow of sound; and, anyhow, the front of the mouth, with reclining teeth and interlocking canines, is all wrong for modulating the sound. In times past, devoted efforts were made to get such apes to say a few words like "mama," "papa," or "cup." Only tortured strangling sounds emerged.

This kind of effort was abandoned, but a great deal of other work with apes has been done. Not only can chimpanzees understand much spoken language (in experiments using headphones in isolation to prevent the chimps from simply watching people for unconscious clues), but they are also good at using standard sign language and at making statements by pointing to conventionalized ideograms on a board. Many scholars are convinced that apes command simple grammar and syntax — the rules fundamental to our language — and also have the ability to handle symbols internally.

But readily transferring ideas from one individual to another — real communication — is something else. At Yale, long a hotbed of chimpanzee education, psychologists set up a contraption to test cooperation and coordination between animals. A box on rails, containing a reward of bananas, could be drawn within reach of two chimps pulling simultaneously on ropes attached to either side of the box; the mechanism would jam if one rope was pulled alone. Chimps quickly and easily mastered the problem. So an ape, feeling the desire for a mid-morning banana, could solicit any colleague who knew the ropes to help him work the machine; a wink and a nod would convey the idea. But let him try this on an uninitiated comrade, out of sight of the works. Utter frustration. The value of communication could be shown in no better way. As Jane Goodall has put it, a chimpanzee or an australopithecine can have a bright idea, but if the idea cannot be talked about it cannot grow.

Students of linguistics, not dismayed by the lack of fossil speech, have been able to develop some suggestive theory based on their increasing understanding of the way language operates. Language puts concrete and abstract things into words. Ideas become encoded in language and thus structure language itself, which then is able to classify and relate things out of experience. Most mammals, and certainly chimpanzees, know their mothers, but there is no sign that they can generalize to mothers as a class, referring not only to "my mother" but also to "Pete's mother." That would be an elementary level; it would take a very much higher stage of reference before a hominid could say: "Watch out for Pete's mother; she'll stick her thumb in your eye."

As culture began to have more regularities, that is, more kinds of form, language had more things to be regular about. Thus, linguists are entirely willing to see steps of advance in language as accompanied by steps of advance in the mental equipment necessary for them. We know from archaeology that there were tools, then combined tools, and then tools to make tools; and linguistic theorists are happy to see these as symbols of increasing complexity needing reflection and support in language. Evolution of intelligence in such terms is easy to comprehend. Thus, we may certainly suspect rudimentary language — very rudimentary — in *Homo habilis*.

Furthermore, in the hominids some physical blocks to speech were progressively removed. With erect posture of the head, the larynx retreated down the throat, allowing a sound-producing tube above it. The increasingly vertical front teeth provided a hollow chamber better suited to allow the whole mouth to manage consonants and vowels. All signs indicate that language, general culture, and tools moved forward, obviously in concert with brain size. The brain presumably was dragged along the evolutionary trail, responding to ever greater skills of speaking and handling. It did not, as Elliot Smith once said, "lead the way."

Chapter 9

Homo Erectus

Before long, about 2 million years ago, *Homo* arrived at a new stage that lasted a long time with little change. Here were people of almost modern size who had an essentially modern skeleton — not entirely modern, however, being heavier and lacking certain refinements of ours. Also, the braincase was beginning at last to dominate the face.

These long-lasting people had some features of the skull that set them off from *Homo habilis* on the one hand and from ourselves on the other.[1] The brain had become bigger, generally ranging from 900 to 1,000 cc, although these are not limits. It was housed in a heavy, low, thick-walled skull unlike that of the australopithecines or *Homo habilis*. This skull shape may look more primitive. But perhaps growth in general size and robustness had outstripped the modest brain enlargement so that, now, the container was larger, relatively, than the thing contained and therefore looked more primitive. In an earlier image, we saw a baseball-sized brain inserted into a relatively large hominid skull and gradually expanding; at this point, we might say, the head skeleton briefly caught up and surpassed the brain, as both grew in concert with a larger body.

At any rate the cranium was low, widest just at the ears, and wide across the back. The forehead was flat to very flat, with a heavy torus, or bar of bone, right across the front over the eyes, leaving the forehead itself pinched in behind it. The profile was low and at the rear was sharply angled, with a mounded ridge running across the upper limit of the neck muscles where they were attached to the

---

1. From this point on, "*Homo habilis*" should be understood to refer to early *Homo*, including both proposed species, *H. habilis* and *H. rudolfensis*.

skull. All of this sounds brutish, but it was as much a matter of robustness as of primitiveness.

This was *Homo erectus*. It constitutes a stage, fully human but far from fully modern and far from fully known. It will loom large in much of what follows. Most people view it as a good, distinct species in our evolution, now extinct; some dispute this view, as we shall see. Trying to comprehend *Homo erectus* is of first importance for the whole story, and for understanding new discoveries as they come along. To quote Michael Day, "the taxon *Homo erectus* is under intense debate in terms of its geographic range, its temporal range, its origin and its evolutionary fate." That hardly overstates things, and it was written before some of the latest new facts.

Lately, for example, the time for *Homo erectus* has been stretched at both ends, with surprising effects on interpretations. Things seemed, not long ago, to be simpler than they probably were: appearance of the new species in Africa and, after some delay, a spread through the tropics to the east, between a million and half a million years ago. That is what the first known Fossil Men (see page 2) seemed to tell us. And in fact the significant thing is seeing human beings beginning, just then, to spread through the rest of the world. But new finds are giving us surprises and puzzles, especially as to the times of events. Let us look at the simpler picture we knew not long ago, and then at the new complications.

In 1984 Richard Leakey and his team extended their prospecting at Lake Turkana to its northwest side. Just as they thought they had finished with a particular locality, the most gifted and experienced of Leakey's associates, Kamoya Kimeu, spotted a skull fragment, and a little careful soil-lifting turned into a lot more. By the time the team had finished excavating, screening, and washing, they had recovered a fairly complete skeleton, lacking only most of the hands and feet. The parts had been dispersed over more than twenty feet after death and during sedimentation; and a small sapling was growing at the skull and disturbing it, another sign of how nature can confound the hopes of anthropologists. This find was spectacular; no early human skeleton even approaching such completeness had ever been recovered; in this respect, Lucy was the runner-up.

The skeleton is catalogued as "West Turkana 15000" (or WT 15000) from the site called Nariokotome. It belonged to a male about twelve years old or younger, with robust limbs. So complete is the skeleton that the boy's height can be estimated at about five feet six inches; he would have been over six feet tall had he lived.

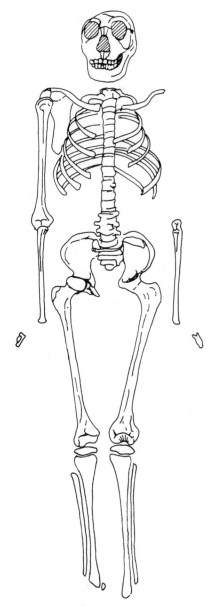

Fig. 31. The Nariokotome Skeleton

This shows the completeness of the skeleton, allowing the probable height of this youth to be estimated at five feet six inches, and his eventual height at well over six feet. His adolescent age is revealed by the fact that the epiphyses of the long bones were not yet fused with their shafts, visible especially in the lower leg. In this early *Homo erectus* the pelvis and limbs are much more modern than those in *Australopithecus*.

That is to say, he would have had the skeleton of a modern six-footer, but he would have lacked a couple of inches at the top because high skulls like ours were still to come.

He does not stand alone; further bones and skulls have been found at Olduvai and elsewhere. Of these, the two best skulls are probably female and are in excellent shape; they are denoted ER 3733 and ER 3883. ("ER" stands for "East Rudolph," the old name of Lake Turkana.) They are well dated at about 1.7 million B.P., slightly younger than the *Homo habilis* remains.[2] They were found before the discovery of WT 15000 and were promptly recognized as *Homo erectus* because of their evident resemblance to charter members of that species in China.

In Africa, as elsewhere, the species has a long history. At Olduvai Gorge a skull cap known as Olduvai Hominid 9 is an obvious member, at an age of about 1.2 million years. Still younger — only about 700,000 years old, or nearly a million years younger than WT 15000 — are a heavy hipbone and thighbone found together at the Gorge (the owner: Olduvai Hominid 28). In spite of his robust bones, OH 28 was not actually a large person. The difference here from modern bones is minor: it lies mainly in some improved design seen in our own hipbones for weight-bearing, a function that in the Olduvai people was served by simple bone thickness. The OH 28 hip resembles that of the much earlier WT 15000 skeleton.

A number of further jaws and other fragmentary specimens from Olduvai are assigned to *Homo erectus* but have not done much to enlarge the picture. Similar people were evidently present in Northwest Africa at the same general time, to judge from three lower jaws and a piece of skullcap recovered at a quarry at Ternifine[3] in Algeria. Finally, in South Africa some fragmentary skull and jaw pieces, which have been identified as *H. erectus* (or possibly *H. habilis*), were found at Swartkrans, along with numerous parts of the local robust australopithecine. Like the evidence in East Africa,

---

2. A couple of hipbones, found at different places, might be those of *H. erectus* and are believed to have an age of nearly two million years. This evidence is not positive enough to date the whole species but, like so much else, will bear watching.

3. Now Tighenif.

these finds suggest robusts making a last stand at a time earlier than a million years ago, well after *Homo* had arrived and the gracile australopithecines were gone.

Now we pause for a note on classification. The description of the robust character of *Homo erectus* given above applies to the Olduvai specimens such as OH 9 and to the Asiatic specimens that follow. But some older skulls, such as those from Koobi Fora (ER 3733, ER 3883) and the skull of the Nariokotome youth, are somewhat less thick and otherwise more delicate, with a few other traits in which they are also closer to *Homo sapiens*. For these reasons the exacting Bernard Wood separates these older specimens, and some others, from *Homo erectus* as belonging to a distinct but closely

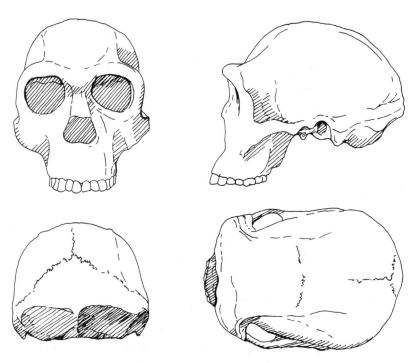

Fig. 32. *Homo erectus* Skull ER 3733
This skull from Lake Turkana is dated at 1.6 million years. Comparison with *Homo habilis* skulls shows the smaller face and higher, rounder braincase. Because it is less robust and has other differences from later *Homo erectus*, some writers recognize it as an immediately parental species, *Homo ergaster*, which would also include the Nariokotome specimen.

Getting Here

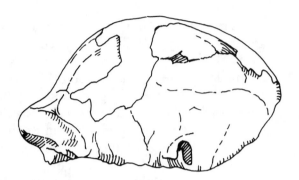

Fig. 33. Olduvai Hominid 9
This partial skull is very thick, with massive brow ridges. In general shape it resembles the earliest Javanese skulls. Thus it looks more primitive than the earlier ER 3733, but is more typical of *Homo erectus*.

allied species, for which he accepts a previously proposed name, *Homo ergaster*.

He is not alone. For a number of workers this possible species expresses a distinction in apparent primitiveness at this very early stage, and placing these early Africans both early in time and perhaps closer to a possible main line to later hominids are propositions that the future will have to take note of. It is not a matter of great concern here.

The Asiatics

The Java Man was the first "early man" to be discovered; a century passed before his cousin WT 15000 was found at Nariokotome on Lake Turkana. The discovery of Java Man was the achievement of a remarkable young Dutch anatomist, Eugène Dubois. As his professor's assistant in Amsterdam in the 1800s, he became disenchanted with studies of the larynx of whales and more deeply intrigued with the possibility of finding human ancestors. In those heady days following Darwin, nobody knew what to expect. The only "antediluvians" that had been found were Neanderthals, and they seemed quite human. Darwin's disciple, the German naturalist Ernst Haeckel, had drawn up a family tree, confidently inserting an imaginary missing link between ape and man that he named *Pithecanthropus alalus* ("speechless ape-man"). The problem was where to find the ape-man.

Eugène Dubois as a young army medical officer going to Indonesia in 1887.

Dubois concluded, for many good reasons, that the Dutch East Indies (of those days) would be a good place to look. He quit his job, joined the army, got himself posted as a medical officer, and left for Sumatra late in 1887. He had no luck getting official support for fossil-hunting and had to do his early cave-searching on his own time.

However, in a few months he had written and sent home an article arguing the paleontological potential of the Indies. In that primitive period, a paper for a scientific journal did not have to be reviewed, refereed, recast, rewritten, and resubmitted. The article got him immediate attention in Holland, and within eleven months, including time for sea post, Dubois was authorized to do research in paleontology and given two engineering assistants and a corps of convict labor to do it with.

Pithecanthropus erectus

With no good fortune in Sumatra, Dubois moved to Java and tried various sites, especially Trinil on the Solo River. Encouraged by finds of animal fossils, his work force moved large amounts of the bone bed at that place. In late 1891 they found the celebrated skullcap, and next August they recovered the thighbone of a fully erect human being lying forty-five feet away. The skull, with its small brain size of about 900 cc, had thick brow ridges and a sharply angled rear where the neck muscles had been attached: clearly what Dubois and the world were looking for. In his formal description in 1894, Dubois took Haeckel's suggestion and named the fossil *Pithecanthropus erectus*. He insisted, then and for the rest of his life, that this species was neither ape nor man but a true transitional missing link (*Übergangsform*). It seemed impossible in those days, before the finding of *Australopithecus*, to consider as really "human" anything as primitive as the Java fossil. Dubois went back to Holland and exhibited his find widely. That it was debatable and controversial was the only thing the anatomists and anthropologists agreed on. Some thought it was human, some thought it an ape; everyone had his special shade of interpretation. What galled Dubois was that no one agreed with his own particular view. In annoyance he eventually took the skull home to his house in Haarlem and concealed it under the dining room floor, where it stayed for many a year.

But the discovery itself was truly a story of pluck and luck. The pluck was Dubois's determination and good sense. The luck was extraordinary: the find should not have been made at all. In the first place, the skullcap is the only cranial evidence that ever came to light at Trinil, in spite of the large amount of digging done by Dubois's workers and the still larger amount done in 1907 by a German expedition. In the second place, it now appears that the thighbone probably came from a later deposit than the skull and is not, as was originally thought, likely to be part of the same individual. This is actually comforting, because other thighbones of *Homo erectus*, such as those found at Olduvai, are heavier or otherwise distinct from the Trinil bone, which is modern in every particular. So in fact *Pithecanthropus* told the world a true story, but with a strangely isolated skull and a false thigh.

A true story, because many more good skulls and pieces of the Java Man have been found in other parts of Java, right down to the

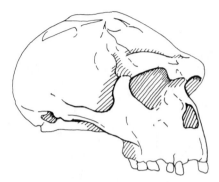

Fig. 34. The Best Java Man Skull
This specimen is Sangiran 17, found in 1969, a member of the later group, and the only one in which braincase and face are present together.

present. In 1937 Ralph von Koenigswald made the first of these later finds, most of which came from around the Sangiran dome (pronounced approximately Sang-NYEERan) not far from Trinil. The dome is the result of a volcanic push-up of a piece of landscape: when this bump weathered down, the oldest, deepest beds were exposed near the center, with progressively later deposits appearing further out. Today, every Indonesian farmer in the region knows what to look for, so the number of finds has been gratifying. Unfortunately, no currently available methods of dating can be applied, but the middle-level beds have been thought to be about 700,000 years old, and the earliest fossils, which are also the heaviest and most primitive, were dated at a million years and more.

The original *Pithecanthropus* has lost his thighbone and his name as well. At the time, awarding him a new genus was the least one could do. But now that all the less advanced African fossils, as far back as Lucy, have been admitted to the hominid family, the scope of the whole family can be seen better and it has been recognized that *Pithecanthropus* should be put in our own genus, *Homo*, though as a separate species. Thus, by the rules he loses his genus name but, as a distinct species he must have his original species name and so he becomes *Homo erectus*.[4]

---

4. To labor a point about naming, this has to be *erectus* because Dubois originally selected that name, not because this was the first erect hominid.

Fig. 35. Java Man in Life
A restoration of a male *"Pithecanthropus"* from a robust early specimen.

The Solo People

In 1931 and 1932 a further group of fossils, apparently later in time than those above, was unearthed (partly by von Koenigswald) at Ngandong, another site along the Solo River. This was a sample of fifteen skulls or skull parts but no face parts to speak of, and without a single tooth.[5] These fossils are immensely heavy and large — more so than the bulk of the Sangiran remains.

The Solo people were evidently robust in general, as is attested by the only other remains found, two shinbones. (The robustness of these bones reinforces the idea that the Trinil thighbone, not being robust, is actually a stray from a different time level.) The Solo skulls are homogeneous in form, with massive brows and thick walls, having their greatest breadth at the ear level and with heavy crests at the back for the neck muscles. Their brains were larger than those of earlier Javanese, averaging 1,050 cc versus about 900 cc earlier. But these Solo people hardly seem more progressive than their

---

5. Tools were found in the terrace, but during World War II the Japanese military occupied the museum in Jakarta and the contents were hopelessly confused.

predecessors, and there are strong arguments for placing them in *Homo erectus*, as a bulkier version of their Sangiran neighbors. The most progressive of the latter are much like the Solo folk in shape and detail, if less massive.

Extremely important, the date of the Solo people had been a dilemma. They had been placed anywhere between 300,000 and 900,000 B.P., more or less by guesswork. This is rather close on the heels of the Sangiran hominids, from whom they differ somewhat cranially. But they have now been convincingly given date estimates of not more than 53,000 B.P. or less — mostly less, down to 27,000 B.P. This raises the much more serious question of overlap with modern people. This is a prime example of new information having a wide effect, and it is causing anguish in some quarters. More later.

Chinese Version

In this same period *Homo erectus* surfaced in North China, forty miles west of Beijing. Compared with the hullabaloo attending the Trinil find, his Chinese entrance was downright stately. He was expected, he was sought for systematically, and he was welcomed with open arms. By the 1920s the anthropological mind had been prepared to accept such low forms as human beings — even while the profession was trying to strangle *Australopithecus* in the cradle!

A Swedish mining adviser turned paleontologist, Johan Andersson, had his attention drawn to fossil-bearing deposits near the village of Zhoukoudian, just where the Western Hills begin; he in turn brought others there: the American Walter Granger and the

Fig. 36. The Sangiran Dome in Cross Section
A diagrammatic cross section of the Sangiran dome, about three miles across. The edges (*1*) are beds laid down, before the uplift, believed to date to the beginning of the Upper Pleistocene, about 120,000 B.P. The next older beds (*2*) are believed to coincide with the Middle Pleistocene, perhaps 780,000 to 120,000 B.P. It is from these beds that most of the Javanese *Homo erectus* fossils have come. The oldest deposits (*3*) have yielded a few older remains.

Austrian Otto Zdansky. In 1921, among animal fossils of a promising age, Andersson noticed white quartz flakes, not natural to the deposit but good material for making stone tools. As reported, he said to Zdansky, "Here is primitive man; now all we have to do is find him."[6]

Find him they did, little by little. Among the fossils Zdansky collected, he later recognized two teeth as probably belonging to *Homo*. A new man on the scene was the Canadian Davidson Black, teaching anatomy at Peking Union Medical College, which became the center of the work (helped by Rockefeller support). In 1926 a project was set up between PUMC and the Geological Survey of China specifically to pursue the hunt for early man in the big deposit, a collapsed and filled cave known as Locality 1, at Zhoukoudian.

Now there was no place for Peking Man to hide. He kept his privacy only until the next year, when another tooth, a good specimen, was found. Black studied this one in great detail, along with the better of the other two teeth, and convinced himself of its hominid nature. In what then seemed like a nervy action — staking all on a couple of teeth — he described them fully in print and formally named the owner "*Sinanthropus pekinensis* Black and Zdansky." Thus he shared the honor with the finder of the first tooth.

The gamble was promptly vindicated the next year, when two lower jaws came to light under the partial direction of Pei Wenzhong, then fresh out of Beijing University but afterwards long central to the work. It was he who, late in the next year, 1929, dug out the first actual skullcap, in fine shape.

Peking Man stood revealed, and there was a real sensation around the world. After all, in that year the whole body of known fossils was not what it is today: from all of Asia only the Trinil skull was then known. But now fresh finds at Zhoukoudian came along regularly, until the total of individuals, represented by a tooth, a jaw or a skull, stood at forty-five.

Tragically, Davidson Black died in 1934. Fortunately, his successor, Franz Weidenreich, a German anatomist, was an experienced, clear-headed, and energetic student of human evolution who over the next fifteen years did much to inform the whole field. He

---

6. He said it in Swedish, or possibly German, in somewhat less dramatic form; it gains something in the translation quoted.

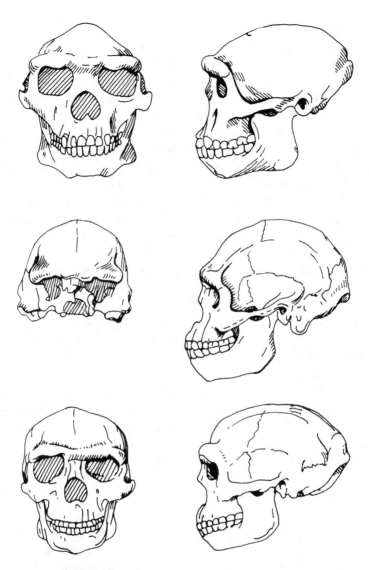

Fig. 37. Far Eastern *Homo erectus*

Top, Java Man as reconstructed by Weidenreich from parts believed to be from the earlier occupation period. This and the probability that the parts are male explain the marked robusticity.

Middle, Solo Man as reconstructed by Weidenreich from one of the most complete skulls; the face and lower jaw are conjectural.

Bottom, Peking Man as reconstructed by Weidenreich from various parts. It is meant to represent a female; this, the later time period, and the larger brain are reflected in the difference from the reconstruction of Java Man. All three groups share much anatomical detail.

arrived to take over at Peking Union Medical College while the work was in full swing. In the best month of all, November 1936, three fine skulls were uncovered by Jia Lanpo, then in charge of digging. Weidenreich in Beijing was sent news of the first of these, and rushed out to Zhoukoudian in such a state of excitement that he began the day by putting his trousers on inside out.

That was the apex, and things went downhill from then on. Work had to be stopped in the late 1930s because the war with Japan had made the countryside dangerous. In 1941 Weidenreich, having studied all the material in great detail, decided that he could do little more in China and left for the hospitality of the American Museum of Natural History in New York. Later that year, the Chinese authorities thought to keep the fossils out of Japanese hands by sending them to America. They were packed carefully in two wooden crates and turned over to the American Embassy for transport by the departing U.S. Marine detachment. This was put in motion a few days before Pearl Harbor. The Marines were captured by the Japanese, and the bones have never been seen since. What happened is simply unknown. The most intensive investigations over fifty years, by Chinese, Japanese, and Americans have turned up nothing.

The only comfort is the work of Weidenreich. It is clear that he was probably the best person, worldwide, to describe and evaluate the Peking Man fossils. In New York, still ignorant of the fate of the fossils, he published his exacting and voluminous studies and descriptions of the material, recording it in this way for posterity.

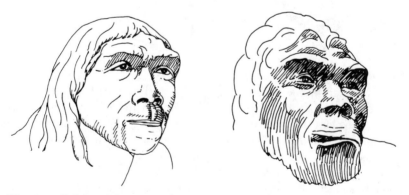

Fig. 38. Peking Man in Life
Restorations of her and his probable appearance.

Franz Weidenreich and Ralph von Koenigswald contemplating one of the Solo skulls, at the American Museum of Natural History in New York in 1946. Von Koenigswald, after spending the war in a Japanese prison camp, came to New York with all the Javanese fossils (which have now been restored to Indonesia).

He had also arranged, before leaving Beijing, to have a technician, Hu Weizhing, make a number of superb casts of a number of the skulls, the likes of which have never been equaled. On the downside are all the further studies, using many new techniques, that now can never be made.[7]

More digging has been done at Zhoukoudian in recent years — a great deal of the cave deposit still remains — but this has not

---

7. We face the same gloomy result when collections of bones are turned over by museums for tribal disposal. However ethically correct, as we must suppose it to be in the typical case, all future investigation is forever precluded. There is an unfortunate view that, if such material has been studied once, it has no further scientific value. Nothing could be further from the truth.

been very fruitful. The main result so far has been retrieving most
of a skull of which part was found before 1941. A similar skull was
recovered at a new site, Hexian; and another, rather crushed skull,
and a lower jaw, at places in Lantian county (Shaanxi province), to
the south. All this adds up to a picture of Chinese *Homo erectus*.
(Black's genus name, *Sinanthropus*, has gone the way of *Pithecanthropus*, as signaling an unwarranted degree of distinction. Today's
accepted designations instead recognize two subspecies, a Javanese
Homo erectus erectus and a Chinese *Homo erectus pekinensis*.)

The main cranial features are the same for Java and China: a
thick, low skull with a heavy, single browridge and an angled back,
or occiput, with a bony torus there also. Weidenreich distinguished
minor differences in the Chinese form: not quite so flat a forehead
but a slightly flatter region just behind the vertex; and, perhaps most
important, a somewhat larger brain. All in all, the Chinese population seems a little less robust (certainly when compared to
Ngandong), and the fragments of the rest of the skeleton convey the
same suggestion. No more than a suggestion, however, because
bones other than skulls or teeth have been extremely scanty.

Time figures in this variation. Long study of the deep cave
deposits in Locality 1 at Zhoukoudian suggests an occupation from
about 500,000 to 230,000 years ago; these are the currently agreed
dates. These are on the whole later than the main Javanese group
from Sangiran and Trinil, and thus indicate that the known Javanese
lot is the earlier, as cranial form might suggest. Within China, the
Lantian skull and jaw are earlier than Zhoukoudian and somewhat
more primitive; the Hexian skull falls at the end of the Zhoukoudian
time zone and seems slightly larger and more advanced. But none
of these specimens violates the apparent anatomical limits of *Homo
erectus* as we know them now.

Time Marches Back

That was the picture as it looked not long ago: *Homo erectus*
arises in Africa (from a somewhat lighter, less developed form, *Homo
ergaster*), spreading into Eastern Asia about a million years ago. No
signs of either of them in Europe. But just lately, better methods of
dating and some new fragmentary finds have led to a lot of head-
scratching, in which the reader is cordially invited to join.

Take Java. On the one hand are the very late dates, on the order
of 50,000, for the Solo people. On the other hand are signs of much

greater age for the Sangiran specimens. Strongly supported new dates argue that the oldest and most rugged remains, as previously recognized, are to be dated, not at less than a million years, but at 1.6 to 1.8 million, almost twice as old as previously believed and contemporary with the lightly built Nariokotome boy, WT 15000 from Kenya.

In China, apparent stone tools here and there suggest similarly early ages. But the shocker comes from Longgupo (Dragon Hill) Cave, near Wushan in eastern Szechuan province. Here are tools persuasively dated at almost 2 million years old, along with teeth of the giant ape *Gigantopithecus*, and also a tooth and a jaw fragment with two more teeth of a hominid. And the teeth are more like *Homo habilis* or like the pre-erectus *Homo ergaster* of Africa, not like *Homo erectus* of Asia, east or south.

So some anthropologists are considering the possibility that *Homo erectus*, as we know him, actually developed in Asia from earlier migrants from Africa, perhaps *Homo ergaster,* and that *Homo erectus* as a species is actually confined to Asia. Given, with the new datings, that the early Javan remains are as old as the more primitive Nariokotome lad from Kenya, this is not an outrageous idea. However, here we will continue to treat all as one species, *H. erectus.*

Between the Far East and Africa, a jaw from Dmanisi in the Georgian Republic in the Caucasus is probably 1.9 million years old, another non-African not yet fully understood. It is identified as *H. erectus* by its finders, but whether it could be dropped unnoticed into a collection of *erectus* jaws is not clear.

A not very satisfactory cranial vault from Narmada in India is accepted as *H. erectus,* and is the only suggestion of the presence of early people from that large area.

And at last, pieces of a skull cap from Ceprano in Italy are enough to argue firmly that *Homo erectus* had in fact arrived in Europe by perhaps 800,000 years ago. Tools were found along with this person, equivalent in form to the advanced Oldowan of Kenya. At Isernia La Pineta in Italy copious tools of recognizable kinds are dated at 730,000 B.P. Still others have been reported at 500,000 years from Siberia, which seems to be a very early northern penetration.

Just now, we really have little idea as to the actual movements of these still primitive human beings, moving as very small groups into new countrysides, facing unfamiliar conditions and changes. The limits seem to have been tropical for a long time before more temperate places, like Europe or even China, were penetrated. Susan Cachel and J. W. K. Harris (at Rutgers University) have an appealing view. They suggest that *Homo erectus* was a "weed species," meaning that, regardless of tools, this kind of human had new resources of adaptation encouraging its success in new, changing, or disrupted environments. This would not favor typical primates — except, they point out, the highly successful and varied macaques, who alone among monkeys have flourished in temperate places like Japan, central China, and Tertiary England. But we are facing changed ideas from new finds, as always, and must try to envisage a world of movements of small tribes adapting their physiques and their simple tools as they cope with new homes outside of Africa.

Chapter 10

Some Housekeeping: Ice, Tools, Names, Lineages

A sense of the world *Homo* was living in, and of the ways he was coping with it, is essential to knowing more about his evolution. Both kinds of study — climate and tool use — become less difficult to trace as we approach our own time but also call for more fine-grained analysis. Both tasks are demanding at any point. We are beginning to discern a great deal about past climate: its probable role, for example, in setting the stage for hominid emergence in East Africa. In fact, we may expect to see such knowledge figure much more prominently in study and discussion.

The Pleistocene Epoch

Back in the 19th century it was realized that there had been an "Ice Age" — that there had been extensive glaciers on the land. Early in the 20th century a couple of Germans named Penck and Bruckner recognized four successive advances of the Alpine glaciers, with intervening retreats. They named the glacial stages Günz, Mindel, Riss, and Würm for little river valleys in southern Germany where they had worked the evidence out. But the important continental ice sheets, like those today in Greenland or Antarctica, lay nearer the North Pole. North America had the greatest ice bodies of all. Here, in the most recent stage, there were two such bodies: one based on the western mountains, the other a huge sheet spreading out of the eastern arctic, coming down over the upper Middle West and New England and pushing rocky debris out to sea to create Long Island, Martha's Vineyard, and Nantucket.

So impressive was this realization that, for a time, geologists tried to cram everything into the fourfold Alpine scheme, making the American and European cycles correspond with one another and also postulating rainy phases in Africa to produce a wet Sahara such as did indeed exist from time to time. All this nice correspondence is now rejected, although the last major cold cycles did correspond generally and now serve as real and vital guides to events. With the known complexities, there is not now any hard-and-fast scale, although climate is still the main guide. Workers fall back on a few kinds of index that may be determined worldwide, such as paleo-magnetic reversals and ocean temperature sequences, the latter traced by oxygen isotopes. These still leave a detailed chronology short of the mark.

It is now clear that the number of glacial advances was much greater than the Alpine four (the Last Glaciation of Europe alone actually comprised at least three) and that the "Ice Age" did not have a neat beginning and end. In fact, it is generally accepted that the glaciers will return in due course. The causes and the controls are not well understood. The distribution of received sunlight is affected by long-term wobbles in the earth's spin axis, combined with the shifting of summer and winter along its orbit, which is not circular. Major ocean currents exchanging water between the tropics and the Arctic are also implicated. And uplifting of main regions of Asia and America is believed to have brought about a cooling of the atmo-sphere. In any case, ever since the Eocene the earth has experienced a gradual, irregular cooling. In the Miocene the ice sheets of Antarctica took hold, locking up large volumes of sea water, so that dropping sea levels exposed a lot of extra land around the other continents.

Thus, in addition to affecting everyday life and game animals, the changes in ice, seas, and climate had much to do with where people could or could not go. With lowered seas, the first American Indians could walk from Asia to the New World across a broad plain. Contrariwise, with the water level back up again, we today can no longer stroll on the bottom of the shallow North Sea although, from time to time, fishermen haul up from the sea floor the stone tools of our ancestors who walked and lived there.

Originally, geologists recognized the Pleistocene and its divi-sions, as in earlier periods, by the successive kinds of animals present. For example, there was a flowering of elephants in the later Miocene. Taken to herald the Pleistocene was the coming of the modern kinds of elephants (including mammoths), with their complexly folded molar teeth. (Their older cousins, the mastodon of the north and the stegodon of the Far East had teeth with quite different shapes.)

Lower Pleistocene. Began about 1.7 million years ago. If you want an index fossil, *Homo erectus* could serve, but only for Africa or Asia. This coincides with the end of a "normal" interruption (the "Olduvai Event") during the last main period of reversed polarity.

Middle Pleistocene. Began about 780,000 years ago, at the start of the present period of normal magnetic polarity. This has some correspondence with the beginning of the major glaciations of the Old World but not those of America. *Homo erectus* was giving way to later forms in the west but not in Asia.

Upper Pleistocene. Began about 130,000 years ago, as the next-to-last glacial phase gave way to the interglacial phase before the Würm. It is here that we see the first possible glimmerings of the modern kind of humanity.

Divisions of the Pleistocene

Another animal new to the Old World was our one-toed horse (and zebra and donkey), who came to Asia out of North America (evolution's primary horse factory), spreading rapidly to Africa and Europe.[1]

This use of animal markers seemed like the best way of making a uniform framework. But it has turned out to be rather untidy and has given way to paleomagnetic dating (compass polarity reversals), which makes a worldwide time frame. The table gives the currently accepted divisions.

The March of Tools

Stone tools appear, and show progress over time, but only loosely in concert with hominid progress. They do not shed much light directly on human evolution as we know it. Tools are, nonetheless,

---

1. An obvious question: why were the Americas horseless in later times, before the Spanish brought them afresh from Europe? Evidently, because the earliest Indians hunted them to extinction along with other large mammals.

of first importance even for the indirect light. We find far more tools than bones, and tools tell us much about ancient activity.

As we have seen, the earliest and crudest samples of chipped pebbles are dated at almost 2.5 million B.P., and we have every reason to think they were the work of *Homo habilis*. Shortly, as with tools found with the 2.3 million-year-old Hadar jaw, they were being made to more of a pattern and so they are more recognizable. In Africa this is called the Oldowan stage of culture. The pattern is not much: a pebble, perhaps half the size of your fist, was held in one hand while a hammerstone in the other was used to knock off a couple of flakes to make a sharper edge on the pebble. This edge was not very sharp but, on the other hand, it was strong enough to be useful. In addition to the pebble tool itself, the small, sharp flakes struck off must have been put to many simple cutting uses.

A major principle here: the edge of a broken flake of stone may be very sharp, and in fact too sharp. It may break down easily when used in heavy cutting or scraping, say, of an animal hide, losing its usefulness. This is the point of all the flaking and of much of the progressively more refined kinds of secondary trimming right up to the advent of metal tools.

An important addition to such pebbles and flakes was the hand axe, index tool of the Acheulean culture tradition. This was a flattish core flaked along both sides of its edges, typically oval in outline with a point at one end. The hand axe was immensely long-lived in Africa and Europe, slowly reaching a high symmetry of form and elegance of shape as time went on and as more fine-grained stone was used. No one knows its main use. It was probably an all-purpose instrument, heavy enough for woodworking as well as for butchering. It would hardly be useful for hunting or fighting; its forerunners and followers also evidently were not fighting tools.

It is called *hand axe* (English), *coup de poing* (French), or *Faustkeil* (German), all of which confesses that we have no idea how *Homo erectus* and his successors might have named it. That would be valuable to know — probably "cut-dig," and doubtless a better name than the above — not only because that would reveal its use but also because it would serve as a clue to the stage of language.

In spite of gradual changes in the hand axe, Arthur Jelinek (University of Arizona) has called the general stagnation in Lower Paleolithic stonework, lasting more than a million years, a picture of "almost unimaginable monotony." We must view this as reflecting a long period of little evolutionary advance in language and

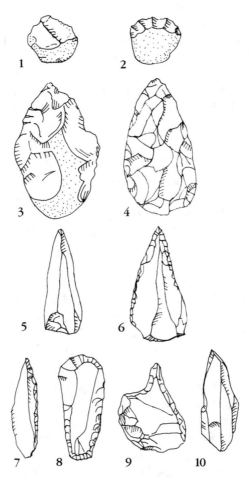

Fig. 39. Progress in Stone Tools During the Paleolithic

1. A small, partly flaked Oldowan pebble tool.
2. A European pebble tool with a straighter flaked edge.
3. An early Acheulean hand axe, ovoid in shape and flaked bifacially.
4. A later Acheulean hand axe, more fully and skillfully flaked.
5. A Mousterian Levallois flake point, showing the upper surface as already prepared by flaking on the core before detaching the flake.
6. A Mousterian point made by further flaking of the edges.
7. A Châtelperron backed blade, i.e., a blade of Upper Paleolithic type, blunted along one edge by flaking for ease in handling as a knife.
8. An Upper Paleolithic blade trimmed by flaking for use as a hide scraper.
9. An Upper Paleolithic blade shaped into a skin borer.
10. A burin or chisel for graving wood or bone, a characteristic Upper Paleolithic tool made by striking flakes downward at the point to make a short sharp edge.

intelligence, which would agree with evidence that there was also a long period of only minor change in brain size and physique, as shown by the human fossils, beginning about 2 million years B.P.

Hand axes are stone cores shaped by removing flakes. The next major phase, the Middle Paleolithic,[2] saw an emphasis on flakes instead. Struck off a larger core, flakes were then shaped by secondary flaking into knives, points, and scrapers; and one important aspect of this was the process by which the core itself was first shaped so as to yield the most suitable flake when this was struck off. This is known as the Levallois technique, a clear advance in complexity. This flake complex particularly describes the Mousterian tradition of Europe, northern Africa, and the Middle East, where flickers of still more advanced techniques occasionally appeared. Finally, the Upper Paleolithic saw a sort of Stone Age industrial revolution in which tools were used to make tools. With great skill, using such implements as bone strikers and pressers, workers were able to produce not flakes but slender blades, which were reshaped into whole tool kits: knives, scrapers, awls, and chisels. The blades were made of fine-grained stone like flint, chert, or obsidian. Also, for the first time, bone was used consistently for many other purposes, especially things like barbed harpoon heads and spear-throwers. This Upper Paleolithic appeared rather abruptly in Western Europe around 35,000 B.P., and somewhat earlier to the east.

That is the general outline. We find no wide correspondences by areas or with known kinds of people. There were important local variations. Beginning with *Homo habilis*, the crudest flaked tools carried on in Africa into the time of *Homo erectus*. Bifacial hand axes did not appear until 1.4 million B.P. or perhaps even 1.2 million B.P., and they continued for a million years. Bifaces appeared in Israel at about 1.4 million B.P., but not in Europe until around the beginning of the Middle Pleistocene. At the moment the Isernia collection at 730,000 B.P. is the one persuasive lot for Europe.

As for the Far East, bifacial tools like hand axes are almost or entirely non-existent, and in any case do not form a tradition like the Acheulean of the west. In Java, only small pebble tools can be associated with *Homo erectus*, while the later Zhoukoudian cave

---

2. This term is intentionally vague, to cover the Mousterian of Europe and the middle East, and the Middle Stone Age of Africa, which share a general character. The period began perhaps 200,000 years ago.

deposits of Peking Man offer only fairly well-trimmed pebbles and flakes, again with no hand axes. It is highly likely that in Asia wood, above all plentiful bamboo, served many of the functions of stone tools in the west. Desmond Clark partook in experiments in China showing that the Asian tools could work anything on bamboo that could be done with a machete, and also that bamboo slivers are efficient knives for butchering meat, a recourse not widely present in the west. Clark has also observed natives of interior New Guinea, who lacked metal tools, butchering pigs with bamboo strips. So the Asiatic core and flake tools, similar from China to Australia, may not actually reflect backwardness in culture as compared to the West. What all this means in the important matter of east-west human contacts is not clear.

Names

As we move toward the present, the matter of naming the fossils becomes important again in understanding anthropological discussion and controversy. Names are always important, but Linnaeus made a great contribution in making them formal: any species is given a genus name and a species name. So a Linnaean name is the same in all languages; and thus a prairie dog and a hound dog do not get confused. Many sensible rules have been added that Linnaeus never heard about. For example, a name once legitimately given to a type of specimen with a specific description cannot be changed. Otherwise, bedlam would result. A few people, recognizing how human *Australopithecus* was, wanted to change his name, which means "ape of the south," to *Australanthropus*, "man of the south." Not allowed, because the name's actual meaning is not the point at all, only the strict identity of the fossil or species being talked about. You can change your own surname but not your social security number, and Linnaean names really do the work of the latter.

Of course, any species can be given a different placement on the basis of better information, which is renaming of a different kind. A chimpanzee is *Pan troglodytes*; a gorilla is *Gorilla gorilla*. Because the two species are so very similar physically, it has often been suggested that they should be placed in the same genus. In that case, the genus name *Gorilla* would have to be dropped, giving way to the older genus name, *Pan*, with gorillas becoming *Pan gorilla*. However correct this may be, it is inconvenient and has not been adopted.

In the similar case, Peking Man was originally carefully and properly named *Sinanthropus pekinensis* on the basis of a couple of teeth, back when distinctions among human fossils were hard to measure. Since then it has been generally accepted that Peking Man is not a different species from Java Man and that both are not generically different from *Homo*. So the Chinese fossils have become *Homo erectus* (subspecies *pekinensis*). Notice that this does not introduce a new name but merely restructures a classification, moving a species to its proper place under an already existing name.

Unfortunately, there used to be a weakness, among some anthropologists not fully conscious of the rules, for tossing off names without a legitimate description or justification, and so the books are full of such vanity items. Later anthropologists have had to clean up after these malefactors by dutifully listing such "synonyms." This situation gives a bad name to still other anthropologists who are merely venturesome, seeking to make finer distinctions, for example, among the *Homo habilis* specimens. Their colleagues may simply say to themselves, "there he goes again," but other readers may find it confusing.

Homo sapiens

Names are not crucial, since they do not substitute for facts. Nevertheless, they organize the facts, and they make statements. Fossils are the picture; a name is a frame.

Here is a good name, and good agreement. All living men, women, and children are one species, *Homo sapiens*. It is a proper animal species by every criterion, especially in that all its populations can interbreed freely. This is unlike, say, horses and donkeys (*Equus caballus* and *Equus asinus*, respectively); these closely related species can breed mules between them, but because mules are infertile, they do not blur the distinction.

Within *Homo sapiens* there is a lot of regional surface difference (race), which will promptly come to haunt us, but the essential bony structure is the same: high, thin skulls, modern limbs, and pelvis. These things differ visibly from the *Homo erectus* remains of China or Africa. In current statistical analyses of measurements, modern skulls the world over are much alike; there are regional distinctions, but separation is not clearcut.

Just about everyone agrees that *Homo erectus* represents a million years or more of human beings. His fossil remains share a number of features, which we have looked at; these features are found in the early Lake Turkana skeleton and skull, and they continue all the way to the Solo people of Java, which is a lot of time.

In the whole history of pre-modern *Homo* we can think of stages: pre-erectus, erectus, and post-erectus. The first, the pre-erectus stage, gives us the still unclear and varied forms put together as *Homo habilis*. The last, the post-erectus stage, our next subject, is represented by numerous finds, ranging from rather primitive to modern in nature but, at least in the early part, they do not make up a very coherent picture. And the specimens are mostly singletons, not groups. By contrast, the long middle stage of *Homo erectus* seems clearer. For one thing, his samples are good. In the 1930s, the fossils jumped from the lonely Trinil skull to good lots in Java (Sangiran and Ngandong) and in China, giving us a solid base of information. We can feel that we know *Homo erectus* when we see him, at least in Asia.

And we certainly know *Homo sapiens*. On the somewhat thin ice of general agreement, all moderns are often designated *Homo sapiens sapiens*. We are not making a vainglorious suggestion that we are extra wise; it is a classificatory statement that we are all members of a single subspecies of the species *Homo sapiens*. This gesture may seem like locking the barn door before the horse is stolen, because it is uncertain how far we should recognize other subspecies of our species. In fact it has been held, especially by the late Carleton Coon, that the major races are in fact geographical subspecies (which, unlike species, can interbreed). But typical animal subspecies, usually showing slight physical distinctions, are normally separated by geographical barriers that human beings readily surmount.

MRE or ROA? A Grade or a Species?

We now know that human beings were present throughout the Old World tropics almost two million years ago, and probably later in Europe. We know that all were then at the stage of *Homo erectus*, or perhaps even more primitive. It is clear that this stage began in Africa, and then moved out.

Today's central argument is over the acknowledgment of a second event. Was there another phase of migration out of Africa? It would have been very much later — less than a tenth the age of the first, not long before 100,000 years ago. And in this case it represented *Homo sapiens*, who had by this time appeared in Africa only.

These people everywhere would have replaced local survivors of the ancient populations.

Either this happened or it did not. That is what the present controversy is about. In one view the ancient populations every-where — in Africa, Europe, China, and Southeast Asia — formed a nexus of tribes, partly interbreeding and partly diverging, which locally evolved from *Homo erectus* into the more recent "racial" peoples of those regions. This view is known as Multiregional Evolution (MRE). It is also known as the Single Species hypothesis, because its adherents hold that the entire Old World population evolved as a genetic but locally varying unit, as above, and that there was no point at which it crossed a species line — in fact, it should be called *Homo sapiens* from start to finish. Some time ago I dubbed this viewpoint the Candelabra model, to give a mental image of populations diverging at the base and then evolving upward in parallel branches.

This hypothesis, in place for over fifty years, was first proposed by Weidenreich and later supported by Carleton Coon. In its present form it was formulated by Milford Wolpoff of the University of Michigan and Alan Thorne of Australian National University. It has other adherents, who subscribe but do not amplify.

In the second, opposing, view any early local peoples deriving from *Homo erectus* were eventually replaced everywhere by the late migrants from Africa. These, and their present descendants, were the new species, *Homo sapiens,* and are seen as too homogeneous to reflect two million years of local diverging evolution. This school is known as that of Replacement, or Recent-Out-of-Africa (ROA). I have earlier called this the Noah's Ark model, picturing the sons of Noah going forth in several directions to repopulate the earth, well after the original creation. Currently the majority opinion, it is modified in different ways by different writers.

Science, we have seen, proceeds by setting up hypotheses to explain facts, as known at any time, and then considers ways to show that the hypotheses are faulty, to see which, if any, survives. So we have, above, two clearly competing models, and one of them, at least, has to be wrong.

Chapter 11

The Problem of the Transitionals

Hypotheses of modern origins are clear enough. The facts are not. There is that problem zone, lasting perhaps six to seven hundred thousand years, during which *Homo erectus* gave way here and there to the large-brained, more lightly built people of today. We know about the Zhoukoudian and Ngandong *erectus* survivors from good samples. But there began to appear more advanced people, and these are known mainly from single finds and fragments, more difficult to interpret. It is a zone of obvious progress after the immensely long stasis of *Homo erectus*.

Regardless of final truth, I am obliged to approach the transition from a Replacement interpretation. This is because Multiregionalism does not admit of significant events and divisions, and in fact smooths them out instead of seeking distinctions.

As I said earlier, we can feel that we know *Homo erectus* when we see him. And we certainly know modern *Homo sapiens*. We have a sort of checklist of his features. Some below the neck have been mentioned as being distinct from *Homo erectus*, especially a lighter pelvis. The skull is also lighter and thinner (with a cranial capacity of about 1,400 cc); it is vertical-sided and high, with a rounded profile at the back and no torus or heavy bar for neck muscles. The lighter, more vertically perched head does not need such support and, instead, needs some support from the front and sides, so that the mastoid processes behind the ear, where the muscles from the breastbone are attached, are well developed. The face is lighter, with smaller teeth, and the pulled-in face and teeth leave a well-marked chin protruding. The brow ridges have become vestigial, exhibiting signs of a division over the eyes between the central and lateral parts.

It might help to think of the story as having two phases. Beginning perhaps 800,000 years ago, there appeared people, not *erectus* but certainly "archaic," not having all the sapiens features above. They have sufficient, if primitive, character of their own, seen in both Africa and Europe, to warrant considering the acceptability of a new species, *Homo heidelbergensis.*

Much later, well within the last 200,000 years, came people generally called *sapiens*, who seem really modern in some cases but not entirely in others, even in the same group. If you met a living "archaic," you would not have him in the house, whereas an early "modern" might simply cause misgivings. Even after the arrival of unquestioned moderns in the second half of that period, skulls and skeletons often showed a kind of robustness not seen today.

Archaics in Africa

We have a sweep of important fossils from Africa and Europe in the early phase, that of "archaics." Africa, persuasively the original home of *Homo erectus,* has yielded his apparent remains at points from Algeria to South Africa, with dates coming down at least to 700,000 B.P. but with no finds after perhaps 400,000 B.P. About then, *Homo erectus* was succeeded by a decidedly archaic set of fossils, reaching at least from the Cape of Good Hope to Ethiopia.

The first-known and the best specimen is the famous "Rhodesian Man," found in 1921. At the town of Kabwe, Zambia (then Northern Rhodesia), there was a prosperous zinc and lead mine named Broken Hill after a similar, more famous mine in Australia. This one was a kopje, or little hill, which had been mostly mined away when a fine skull came to light from what appears to have been a filled cave in the kopje.

In appearance, the Broken Hill skull seems only barely out of the *erectus* class, but it is recognizably different from *H. erectus* as already described. The braincase is long and low with heavy brows, a large protruding face, and very big palate and teeth. However, the sides of the skull are vertical, the bone is thinner than that of *H. erectus*, and the volume of the braincase is greater — just under 1,300 cc. Some other bones were found at the site, not necessarily from the same individual or from the same time. One shin bone is long and robust; a hip bone has a reinforcing pillar above the joint, like the *H. erectus*

Broken Hill, Zambia
Bodo, Ethiopia
Ndutu, Kenya
Hopefield, Cape Province
Florisbad, Orange Free State

Early Post-erectus Finds in Africa

specimen from Olduvai. In this last detail the owner was obviously less than modern.

Comparable African finds since then are the Bodo, Hopefield, and Ndutu fossils. They are all probably early, with Bodo at perhaps 600,000 years, ages being suggested mostly by accompanying animal remains. Unfortunately, such African faunal sequences are not very precise indicators. But Günter Bräuer has shown, partly by mathematical methods, that the skulls are all much alike, plausibly representing an early, erectus-derived stem in southern Africa. The most recent reconstructions support this fully.

According to one good authority, Richard Klein, all the above African finds may be dated at 300,000 to 400,000 BP or more, as suggested for Bodo above.

Archaics in Europe

It has sometimes been proposed, for want of evidence, that *Homo erectus* (or *Homo ergaster*) never got to Europe. Now a fragmented skullcap from Ceprano in Italy is good enough to show that he did. This specimen dates from 700,000 to 800,000 B.P., and was accompanied by crude but convincing chopping tools.

Otherwise, the oldest European bones are the Arago Cave remains from the French Pyrénées, the Petralona skull from Greece,

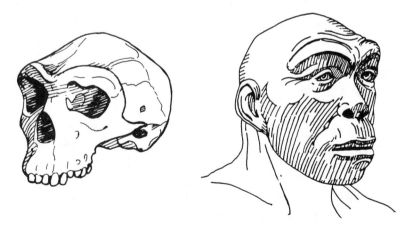

Fig. 40. The Broken Hill Skull and Broken Hill Man in the Flesh
The brows are massive but the braincase is larger and straight-sided compared to that of *Homo erectus*.

a stout tibia of the lower leg from Boxgrove, England, and the Mauer (near Heidelberg) jaw of 1908,[1] the latter being the first really ancient human fossil to be found in Europe. This jaw lay in a deep sand deposit being dug commercially, and has been thought to date anywhere from 400,000 B.P. to considerably earlier.

The Arago cave is believed to have an age in the range of 500,000 to 300,000 B.P. It has yielded a good face and parietal bone as well as two lower jaws. The excavators, Henry and Antoinette de Lumley, think that the skull, with its modest brain size, might be included in *Homo erectus,* though others disagree. Indeed, a well-preserved Arago hip bone has a reinforcing pillar, like *H. erectus* hipbones from Olduvai but also like the Broken Hill example.

The Petralona fossil is dated by consensus between 350,000 and 400,000 B.P. or earlier. Consensus is all we have, because there is not much else by way of evidence, except for vague animal associations. Found in a cave, the skull was not retrieved carefully, and some pretty obvious fairy stories have been circulated about its discovery, which are not worth relating.

Petralona is a fine and important skull, which can be diagnosed as early archaic, just up from *Homo erectus.* In fact, it has a rather strong resemblance to Broken Hill, both as to primitiveness and as to a number of details. Rupert Murrill has inventoried these details, and G. van Vark has used multivariate statistical analyses leading him to conclude that Petralona and Broken Hill (also Arago) form a group distinct from other major fossils. Günter Bräuer of Hamburg has independently done the same thing, placing Petralona especially close to Broken Hill's near relative, the Bodo skull of Ethiopia. Furthermore, different workers have, on occasion, shown that the Heidelberg jaw makes a good fit for the Broken Hill skull.

Somewhat younger are an occipital bone from Vértesszöllös in Hungary and an occipital bone, with further skull parts, from Bilzingsleben in Germany. These people and the Arago lot were all making tools of a near-Acheulean character, including only a few hand axes but many varied small tools of well-developed shapes, far more advanced than the earliest African tool forms. The Vértesszöllös and Bilzingsleben cranial parts are heavily constructed, certainly "archaic" in nature. Finally, from two Acheulean sites in Italy (one

1. This gives the species name to *Homo heidelbergensis* by priority of finding.

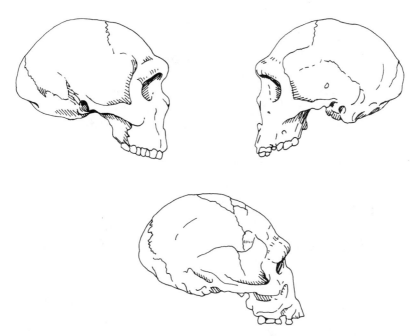

Fig. 41. Post-erectus Europeans: Petralona and Arago

Above left, Petralona. The skull is massive, low, and broad across the base. The brows are large with large air chambers, or sinuses. The cranial capacity is about 1,200 cubic centimeters, not much above *Homo erectus* in general. In all these features, and in its general shape, it strongly resembles the Broken Hill skull (*above right*). J.-J. Hublin finds some elements (cheekbone and nose) in which Petralona approaches Neanderthals somewhat more than does Broken Hill.

Below, Arago, reconstructed from several parts. This is also very archaic, being low, with strong brows and a cranial capacity under 1,200 cubic centimeters.

near Rome, one in the south) a few fragments of skull and thighbones show the same general character as the other early Europeans above.

The implications of all this, especially the skein of Europe-Africa likenesses, are certainly interesting. Günther Bräuer (Hamburg) comes right to the point: he holds that a late erectus stem, at about the beginning of the Middle Pleistocene (700,000 B.P.), gave rise to both populations, north and south. Philip Rightmire and some others agree substantially, with Rightmire persuasively suggesting that the whole group should be recognized as a species *Homo heidelbergensis*.

Notice at once that this species cannot be accommodated by either of the dueling hypotheses described. Not by MRE (no speciation allowed, genetic unity throughout demanded), and not by ROA (not recent in time, not tied to Africa).

However, it is a useful idea, and hypotheses are only hypotheses, not fact. It is more than giving a formal name to the "archaics," because it proposes a real human development and population, provisionally putting a number of the fossils together and facing up to certain problems. Are these early Africans and Europeans indeed a common root and species? Did it emerge in Africa, or did it arise elsewhere, say in western Asia, spreading back to Africa? It is quite possibly contemporaneous with some *Homo erectus* survivors in Africa, and certainly so with survivors in China and Java. This is a first embarrassment for the Multiregional Evolution hypothesis. But it also reminds us how little we know about population sizes, about possible interconnections among such early Middle Pleistocene peoples, and about where *Homo heidelbergensis* might have led.

The Asian Complication

Asia is the problem. Would *H. heidelbergensis* accommodate certain "archaics" of Asia? Dennis Etler, at Berkeley, has reviewed the often fragmentary remains and doubts such a simple solution.

We have seen a certain grudging evolutionary advance in the "Peking Man" population, but neither this nor the Solo people moved convincingly out of *Homo erectus*. The last obviously has a huge overlap in time with both *heidelbergensis* and *sapiens*. We may grant that this was an island population, perhaps isolated in its recency. But notice that the latest date for Zhoukoudian, about 230,000 B.P., has always seemed rather later than would be expected for vintage *H. erectus*. And it is followed by an even later skull from Hexian in the south, still clearly *H. erectus* but a little more advanced — not so pinched in at the sides behind the eyes. The same description fits a pair of large skulls newly found at Yunxian in Hupei Province and dated to at least 350,000 B.P. Etler thinks they suggest both the most recent Zhoukoudian people and "archaics" of the west.

Later, from about 300,000 B.P. on, there are various premodern scraps. A fine skull from Dali, in north central China, is clearly out of the *erectus* range but is certainly "archaic"; it has bulbous brows and a thick braincase, but the braincase is rounded and the face is short and pulled in, more like moderns. Its date is currently thought

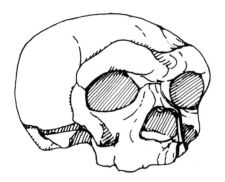

Fig. 42. The Dali Skull
This premodern Chinese skull has a reduced face but a low skull and large continuous brows. It is very different from contemporary pre-Neanderthals of the west, described in the next chapter.

to be about 280,000 B.P. on. A partial skeleton from Jinniushan in the north is given the same age, which, like Dali, would come near to overlapping with *Homo erectus* at Zhoukoudian. But this skull is certainly not *H. erectus*. It is easily seen to be an "archaic,"[2] with a brain size well above the modern minimum; the skull has no particular resemblance to the Dali person. Colin Groves and Marta Lahr consider that both might be accommodated in *Homo heidelbergensis,* which would be an important allocation, though perhaps straining the integrity of that species.

Otherwise, remains are poor or fragmentary. A partial skull from Mapa, date not clear, has marked but not massive brows and a narrowed braincase. Advocates of the Multiregional Evolution view (MRE), including Chinese scholars, profess to see a gradual development toward living Chinese, passing through Jinniushan. Etler sees persistence of traits of the teeth in some specimens from the Zhoukoudian *H. erectus,* and also

> Dali
> Jinniushan
> Mapa
>
> *Post-erectus Chinese*

an early presence of an important pattern, termed sinodonty, which is thought to be characteristic only of much later people of North Asia.

---

2. The Chinese call it *Homo sapiens,* though not in the sense used herein.

In the meanwhile, these fossils are giving slight headaches to proponents of *Homo heidelbergensis*. How are these almost-moderns of East Asia really to be connected with almost-moderns of the west? There is a lot of empty space between. How was it crossed, and when? This particular matter is probably the crucial one in the whole problem; only patience and more fossils will resolve it.

African Breakout

Africa may be more interesting. An important face and forehead from Florisbad in the Orange Free State, South Africa, appears more progressive than the above Asiatics, and is called a "late archaic" by Bräuer, which seems apt. Carleton Coon thought it might be an ancestral Bushman, and in support of this he hoped it would not turn out to be more than a few thousand years old. But it did; it is now dated at 259,000 B.P., which is surprisingly early for its form; it is likened to Ngaloba, below, and to some advancing North Africans, coming soon.

Also more precisely dated than the Chinese remains, at about 130,000 B.P. or the end of the Middle Pleistocene, is a new wave of African finds that are clearly progressive as compared with the earlier group. In fact, with this new wave we arrive at the border of modern *Homo sapiens*. The Ngaloba skull, with face, found by Mary Leakey at Laetoli in Tanzania (site of the ancient footprints), looks like that of a crude modern, with a short, retracted face and a long narrow skull and moderately large brows; but its severest critics have not allowed it to cross the line to full-fledged *Homo sapiens*.

In Ethiopia in 1967, Richard Leakey and his group spotted two crucial specimens in the valley of the Omo River, which flows into the northern end of Lake Turkana. The fossils, designated Omo 1 and Omo 2, were spied at points a kilometer apart but in the same level of river deposits, to which repeated testings of available materials assign a date of 130,000 B.P. (This is beyond the range of radiocarbon dating, and estimates based on the dating of snail shells by other methods are not entirely satisfactory; however, the results have been consistent.)

Eliye Springs, Kenya
Ngaloba (Laetoli 18)
Omo 1
Omo 2

Late Mid-Pleistocene Finds

Omo 2 consists of a braincase of good size but with brows and other features that debar it from the anatomically modern club.

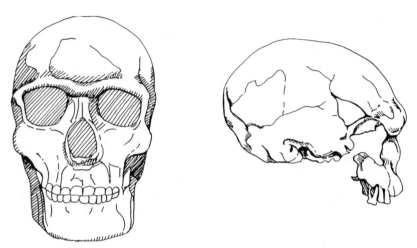

Fig. 43. Two Borderline Skulls, Dated to About 130,000 B.P.
Left, Omo 1. Large, robust, and with strong brow ridges. Although restored from fragments, the skull nevertheless has the character and various special features of moderns. *Right*, Ngaloba (Laetoli Hominid 18), with a short and non-projecting face, still has some archaic traits: marked brows, low forehead, juxta-mastoid crest.

However, like Ngaloba, it is a close approach. Omo 1, on the other hand, is a large partial skeleton with a broken skull and face, which in no respect can be so eliminated. The skull is rugged and cannot be recognized as belonging to any present human population; rather, it is what an earlier, nonspecific example of anatomical moderns might be expected to look like. We have to call it *H. sapiens* on the basis of its known features, and we will bring it up frequently again.

The differences between Omo 1 and Omo 2 are great enough to keep us from comfortably considering them as members of the same level of archaism; nevertheless, they have aspects in common and are not separated in an absolute sense. Each has as good a claim as the other to the date of 130,000 B.P. Perhaps some method of dating will finally separate them somewhat in time, making Omo 1 younger or Omo 2 older; in the meanwhile they suggest that both were in fact close to the time when *Homo sapiens* first appeared in East Africa.

This would certainly be early for moderns. However, discoveries in North Africa, which is shaping up as another important region for anthropologists, suggest similarly early dates. Some finds are apparently as early as the Omo skulls, if not earlier, and stand at a sort of archaic-modern border. These fossils are of great significance, and they will be the first "moderns" we will discuss.

It would be satisfying if we could set out a well-defined time period for the "archaics." But we are dealing with a zone of evolution rather than one of time; it is ragged at both ends, and does not seem to coincide in the various regions. Scholars are working hard on dates, but they are still uncertain for the beginning of these events, and writers do not like to commit themselves beyond something like "later Middle Pleistocene," meaning perhaps after 400,000 B.P. and down to about 130,000 B.P. It is only here, toward the end of the Middle Pleistocene, that some of the fossils are more securely dated.

At any rate, we come here to a crucial phase. Somehow, *Homo heidelbergensis* and the near-moderns above were succeeded, or supplanted, by several kinds of new people.

Three of them were increasingly modern, and we see them at intervals of about 25,000 years. They were: 1) robust virtual moderns in Africa and the Near East, from 100,000 years ago; 2) a separate robust line in Southeast Asia at about 75,000 years ago; and 3) the first fully cooked moderns in Europe, after 50,000 years ago.

First, however, we face another great branch of late humanity, which was separate from all the above: the Neanderthals, of Europe and western Asia. Their history is fairly well known, compared to those others. Because they were looming over the horizon for so long we must pay special attention to them now. As a special point, Neanderthals were never present in Africa, and until late in the story, moderns were not present in Europe.

Chapter 12

The Neanderthals: A Melancholy Story

Two hundred years before Darwin, there lived in Düsseldorf a poet and composer of hymns named Joachim Neumann. He liked to be known by the Greek translation of his name, Neander (New Man), and he also liked to visit a pretty little valley outside the city, through which there flowed a small stream, the Düssel. In time, the people of Düsseldorf named the valley after him, calling it the Neanderthal (Germans of that time still spelled "thal" [valley] with an 'h').

The valley was a narrow gorge then, but in the nineteenth century its sides were quarried away for limestone. Workmen dug out a cave in its flank and found some human bones, which they threw out. Luckily, the proprietor of the place saw the bones and saved some of them: at least the skull top, arm and thigh bones, and part of the pelvis. We wonder how many other ancient bones went unnoticed in those days, just as we may wonder how many treasures at Olduvai wash out and then wash away forever when no Leakey is watching.

The bones were given to J.C. Fuhlrott, a teacher and natural historian in nearby Eberfeld. He was acute enough to note their peculiar features: the bowed limbs and the bulging brows, and he concluded, like some others, that this ancient man was a victim of the Biblical Flood. Fuhlrott turned over the bones to anatomy Professor Schaafhausen in Bonn, where they remain today.

Science was at last ripe for discussions of evolution, as the reception of Darwin's *Origin of Species* shortly demonstrated. But there was then little context of knowledge of time and none of ancient man, and every kind of guess was made about the nature of the Neanderthal person: freak, victim of disease, a Cossack soldier pursuing Napoleon's army, or merely an unusual modern man. The most rational opinion was that of the anatomist William King of Galway, who proposed that a new species of human being had been discovered, which he named *Homo neanderthalensis*.

Over the next century, intensive excavation of caves and rock shelters produced many good Neanderthal skeletons and skulls, whose names are now household words in anthropology. Several of these sites yielded groups of individuals, and numerous more fragmentary parts are known from elsewhere. All are easily recognized as Neanderthals. Their territory was Europe, but extended into western Asia, into Iraq and Uzbekistan, and probably into Siberia.

Germany: *Neanderthal*
Belgium: *Spy*
France: *La Ferrassie, La Quina, La Chapelle-aux-Saints, Le Moustier*
Italy: *Monte Circeo*
Israel: *Amud, Kebara*
Iraq: *Shanidar*
Uzbekistan: *Teshik-Tash*

Principal Neanderthal Specimens of the Last Glacial Phase

The Neanderthal Character

Because of such plentiful remains, the anatomical differences between Neanderthals and ourselves are well known. Brain size is the same in both peoples. The Neanderthal skull is long and low but broad above the ears, so that viewed from behind it has a distinctive round outline. The rear or occiput protrudes in a special "bun-shaped" form but without a heavy crest for neck muscles. Brows are strongly developed but the bone in them is not heavy, and the hollow sinuses inside the brows are large multi-celled chambers. In addition, there are usually present several special features not seen in modern people, for example, a small indentation just above the occipital mound (a supra-iniac fossa) and a small ridge of bone projecting down just inside the mastoid process (a juxta-mastoid crest).

The Neanderthal face is long and projects markedly at the midline, so that there is no angle to the cheekbone, which slopes smoothly back without a depression or canine fossa under the eye opening. The nose is wide and also projecting. A striking feature is the special nature of small bones internal to the nose, and enlarged sinuses on either side of the cavity. These traits were recently

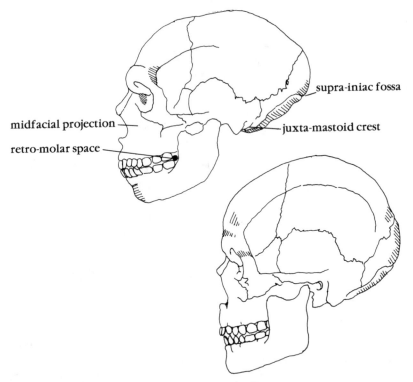

supra-iniac fossa

midfacial projection

juxta-mastoid crest

retro-molar space

Fig. 44. Neanderthal and Modern Skulls

Contrasting features of a Neanderthal (La Chapelle-aux-Saints, above) compared with a modern skull. In general:

1. The vault is long and low, with a projecting, or "bun-shaped," occiput.

2. The face is markedly projecting in the midline. The sides of the cheekbone are flattish, so that its forward border slants toward the rear. The inner wall of the eye-socket and the first molar tooth are placed well forward. By contrast the modern face is vertical and retracted, so that the cheekbone is angled and hollowed in front.

3. The lower jaw is slung forward: there is a sloping or retreating chin, the angle of the jaw is a wide one, and the forward placement of the teeth is reflected in the "retro-molar space," an open space between the last molar tooth and the upright branch of the jaw. In moderns, this branch is more vertical, usually hiding much of the last molar.

4. As special features, there is a small depression, the supra-iniac fossa, just above the bony mound marking the attachment of the neck muscles; the mastoid process behind the ear is small; and just inside it there is a downward ridge, the juxta-mastoid crest, rare in moderns.

Fig. 45. A Restoration of the Neanderthal Head and Torso

detected by J. H. Schwartz at the University of Pittsburgh, and Ian Tattersall at the Natural History Museum, New York, who has written elegantly on Neanderthals. The traits do not appear in other humans, primates, or mammals.

Also striking is the carrying forward of the whole tooth row so that the chin below it is not pointed but slopes back. Connected with this is a "retro-molar space": that is, the teeth are set so far forward that, seen from the side, there is a space between the last molar and the rising branch of the lower jaw; in us, that branch hides most of the last molar.

Why did Neanderthal man develop this highly special face? One suggested explanation is that the force of powerful biting needed a face of such height and size to resist it; but earlier men had bites at least as forceful without developing such a physiognomy. Appealing to many is another kind of possible adaptation: although they made very good stone tools, Neanderthals certainly used their front teeth for more than chewing. Heavy wear obviously resulted from activities like stripping sinews or cleaning animal skins, and signs of such abuse appear in a number of individuals. (Eskimos use their teeth heavily to soften skins.) And so by small-scale evolution this led to the protruding face. I do not much care for this explanation myself.

Finally, Carleton Coon suggested that a cold, damp climate in late glacial times induced the evolution of the big nasal chamber for warming inhaled air. This is certainly a good explanation for the internal nasal structure noted above. And, as Chris Stringer has pointed out, Neanderthals were the first humans known to show signs of cold adaptations. The fact is, however, the Neanderthal face

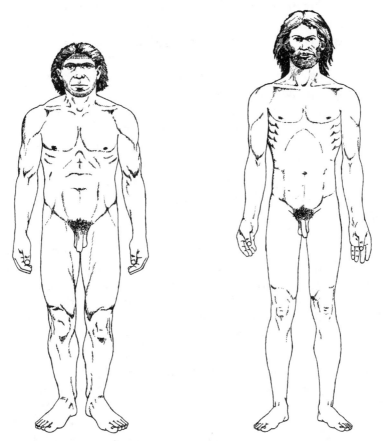

Fig. 46. Neanderthal and Modern Physique
A careful reconstruction of body form in Neanderthal man and his successor Cro Magnon man, who was modern but robust. The Neanderthal is not primitive, but was shorter and heavy in both skeleton and muscle, and relatively somewhat shorter in the lower arm and leg. What do not appear in these external views are the various special differences in the skeleton: the robust and somewhat bowed limb bones, the large joints, details of the pelvis and shoulder blades, and the special muscular effects in hands and feet. For more specific detail, see text.

probably began to develop earlier, during warmer periods. In any case, we know that it is very different.

Neanderthal man is often caricatured as a hunched-over, round-shouldered, long-armed brute, with a stupid expression to boot. This is totally in error. As was noted in the original Neanderthal skeleton, the limb bones are distinctly bowed, and the lower arm and lower leg are short relative to the upper part of the limb, not long. There are now seen to be many minor differences from modern man in the bones of the wrist, ankle, hand, and foot. As Erik Trinkaus has shown, this all reflects a single major pattern, one, as compared with our own, of a heavily muscled body and limbs. Such a form must have been expensive in energy but also one more productive of body heat in a cold climate and one better equipped for close encounters with any kind of animal.

In addition, with their relatively shorter limbs the Neanderthals obey "Allen's rule": among related animals, those in colder parts of a range have shorter limbs and tails, exposing less body surface for any given body mass, and thus reducing heat loss.

So the Neanderthals were a highly evolved people, while also differing from any now living. They cannot be seen simply as men like ourselves who regularly became bulky through life habits, together with some degree of population adaptation to cold, like Eskimos. Rather, their traits bespeak a long-term special evolution, because those traits evidently were genetically deep-seated: they appear repeatedly in the skeletons of young children, who thus are easily recognized as Neanderthals. For one thing, children's milk molar teeth had significantly bulkier pulp chambers and thinner enamel, for whatever reason. And all these traits were a long time in becoming established, as we shall presently see.

To emphasize a point, Neanderthals were not so much archaic as divergent. Archaic they certainly were, in various ways compared to moderns, but as used herein the word applies to more general forms with retentions of early traits, such as appear in pre-moderns, early and late.

Early Neanderthals

Did the Neanderthals derive directly from *Homo heidelbergensis*, as represented by the European fossils named earlier? There is no sign the Neanderthals came from elsewhere. And the fossils next on the scene make plausible Neanderthal ancestors. First is a good undistorted skull from Relingen, Germany, dated only to a warm period between 450,000 and 200,000 years ago; it has the look of a Neanderthal in the early stages of making. Next are the well-known Swanscombe (England) and Steinheim (Germany) skulls of the Great Interglacial Phase, the last but one (say 280,000 B.P.), and the Fontéchevade and Biache pieces of the following glacial phase, the Penultimate Glacial. The Steinheim woman was small in brain and had large, bony brow ridges, but she also had a canine fossa under the eye socket and lacked any of the special facial protrusion of the later Neanderthals. The Swanscombe specimen has only the sides and back of the skull, which are vertical and rounded, respectively,

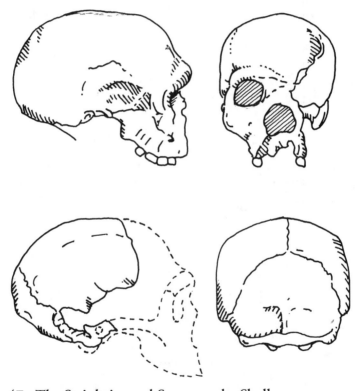

Fig. 47. The Steinheim and Swanscombe Skulls
Above, the Steinheim skull (Germany) is rather primitive and does not show obvious Neanderthal traits. *Below*, the Swanscombe parts (lower Thames valley) look deceptively modern (the face is conjectural), with a moderately high and flat-sided vault, but has a couple of Neanderthal details. Both skulls are obviously post-erectus.

		Europe	Near East
35,000		St. Césaire	
		European Neanderthals	Kebara
75,000	Classics		Amud
		Saccopastore La Chaise Ehringsdorf	Shanidar Skhūl
100,000			(Skhūl, Qafza)
	Proto-Neanderthals	Krapina	Tabūn?
		Fontéchevade	
		Biache Reilingen	
280,000		Swanscombe, Steinheim Atapuerca	Zuttiyeh
400,000	Pre-Neanderthals	Arago, Petralona Bilzingsleben	
		Vértesszöllös	
		Mauer	

Fig. 48. A Time Frame for Neanderthals and Their Forerunners
A time ordering of the important specimens of Europe and the Near East.

so much so that for a while this, like the Steinheim skull, was thought to be a very early representative of anatomical moderns. But lo, the Swanscombe parts also have a juxta-mastoid ridge and a supra-iniac fossa, both of these being telltale Neanderthal features. The Fontéchevade skull has additional furtive signs of this kind, while the Biache outlines are still more positive. Taken all together, these crania suggest Neanderthals in the making. Bits and pieces of equivalent age from other places in Europe say the same, especially jaw fragments with retro-molar spaces, demonstrating the presence of a Neanderthal facial projection in the missing face.

This is persuasive evidence that Neanderthal evolution in Europe can be traced back for at least a quarter of a million years. Such a finding is important in itself, as the best yardstick we now have for the pace of late human evolution in any quarter. Still better are some excellent new skulls from Atapuerca in Spain and Altamura in Italy, which are of a similar age; as known, they are persuasively Neanderthal-looking, although it will take years to free the Italian specimen, embedded in encrustation at its site of finding.

The trend becomes clearer in the Last Interglacial, beginning about 130,000 B.P. Two skulls found at Saccopastore, outside Rome, and a skullcap from Ehringsdorf, Germany, are obvious Neanderthals, though without quite so strongly marked a character as the "classics."[1] This applies, apparently, to the Gibraltar woman[2] of 1848. Finally, the fragments from Krapina in Croatia, now dated to the very end of the Last Interglacial, seem to have all the Neanderthal traits in place. By the early Last Glacial, apparently only classics are known anywhere in the Neanderthal world.

---

1. As a historical note, because of their more restrained expression of Neanderthal characters, these earlier specimens were once termed "progressive" Neanderthals in distinction from the "classics," whence the use of the last term. This led to the suggestion that "progressives" might have been ancestral in Europe to both Neanderthals and moderns, an idea not in keeping with recent evidence.

2. Telling the sex of a skull is a matter of judgment, there being no absolute differences in the sexes, and cannot be determined with certainty in recent crania, let alone earlier people. Female skulls are usually somewhat more lightly constructed, especially in the face and lower jaw, with smaller brow ridges and so on. Diagnosis is greatly helped if some of the skeleton is present, especially the pelvis, although here again there are no absolute criteria.

An End and a Beginning

Now, roughly 40,000 years ago, there took place one of the spectacular events of prehistory: the advent in Europe of moderns — call them the Cro Magnons — and the introduction of the Upper Paleolithic. The people were not the first moderns, only the first in Europe. This new tradition of stone-working, distinguished by the production of thinner blades making possible an array of special new tools, emerged from the Middle Paleolithic somewhere on the periphery of Neanderthal territory. This was not western Europe, where the Upper Paleolithic was a later intruder. But a similar arrival and replacement, some time after 40,000 B.P., is seen in Iraq and in western and central Siberia.

The first full-blown Upper Paleolithic of Europe proper appeared as the important Aurignacian culture, of which the same characteristic tools reached from Spain to Israel. The preceding Mousterian of Europe, well known and manifesting various expert subtraditions, is clearly recognizable as the handiwork of Neanderthals. It contains only a few crude bone tools, no art, and no articles of personal adornment like necklaces. In contrast, the Aurignacian had not only more complex new stone techniques, primarily in blade production, but also bone work, both utilitarian and ornamental. And it was rich in personal decoration in the form of necklaces and other ornaments, also seeing the beginnings of cave art. This is not to say that Neanderthals could not have made such decorative objects, only that they did not do so.

The place of origin of the Aurignacian culture, or its connections with the Near East are not now clear. But, for Europe it first makes its appearance at points in Romania and Bulgaria, certainly earlier than 43,000 B.P.; from the important site of Bacho Kiro, some very fragmentary human remains are definitely not Neanderthalian. The Aurignacian culture was a traveler, reaching northern Spain before 38,000 B.P. and France about 36,000 B.P. And thereby hangs a tale.

Before the Aurignacian completed its occupation of western Europe, it was preceded by a few local curtain-raisers. These were short-lived cultures evincing an Upper Paleolithic character in stonework but in each case also reflecting the local variety of Mousterian culture, unlike the homogeneous Aurignacian. One of these was the Châtelperronian of southwestern France, conventionally classified as early Upper Paleolithic: it did indeed have blade flakes and similar tools and a few objects of art. Usually, the Châtelperronian layers lie immediately below the succeeding Aurignacian deposits; but in at least three places its levels have been found actually interleaved with those of the lowest Aurignacian, showing that the users of each were simultaneously present in Europe, and came to occupy the same sites at different times.

Wherever human bones are found in the Upper Paleolithic they are those of moderns. However, at the site of St. Césaire (Charente-Maritime), a Châtelperronian level lay just below an Aurignacian level, with a skeleton in the Châtelperronian layer. It was that of a young Neanderthal woman, slenderly built; not a transitional form nor with any signs of mixture, but a Neanderthal through and through.

This seems to mean simply that Châtelperronians were Neanderthals who were in contact with arriving Aurignacians, and borrowed some of the new ideas in stone-working. But the Châtelperronian culture disappeared without issue, and so also, as far as there are any signs, did Neanderthals. Various scenarios have been proposed. We do not have to imagine bloody battles or in fact

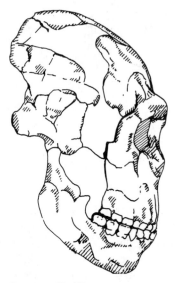

Fig. 49. The St. Césaire Skull
Although that of a non-robust female, it shows fully Neanderthal traits of facial projection, type of brow, and retro-molar space.

any serious conflict: a small erosion of Neanderthals in each generation would have quickly led to extinction. Also, the Upper Paleolithic people were manifestly more numerous than their
predecessors. Another simple explanation is one of vastly different
life styles and hunting capacities, with available game going mostly
to the new people. Certainly the moderns, with their longer limbs
and lighter build, would have been better distance walkers and
runners. When the Boston Marathon is held, no Neanderthals need
apply.

There are other views. One, the "Neanderthal phase" idea of
the Multiregional conviction, sees Neanderthals rapidly evolving *in
situ* into Europeans. Another sees Neanderthals absorbed through

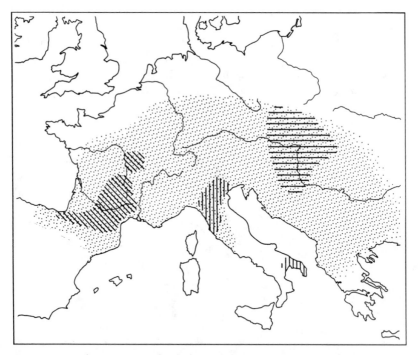

Fig. 50. Early Upper Paleolithic Cultures in Europe
Distribution of the earliest Upper Paleolithic cultures in Europe. Against
the wide reach of the homogeneous Aurignacian are shown the areas of
three more localized cultures: these are Upper Paleolithic in character but
in each case also reflect the nature of the previously existing regional
Mousterian. The Châtelperronian is the most western of these. The picture
suggested is one of local Neanderthal communities whose industries were
modified by the impact of incoming Aurignacian-users.

mixture with the new people. A more sophisticated version of the last holds that the flow of new genes into the Neanderthal population rapidly transformed it in the short space of a few thousand years. I would suggest that the Neanderthal physique and genetic structure were too entrenched for so facile a modification. As for the proposal that Neanderthals evolved into specifically European moderns, there is simply no evidence that I can see.

Few believe that in the far west — as at St. Césaire — there was anything but outright replacement. Some investigators think they see transitional forms from Central European sites, but the material is not very good. Certain Upper Paleolithic skulls, with strong brows and somewhat projecting faces (found at Mladeč and Předmostí, in the Czech and Slovak Republics), seem to some to represent either mixtures or else vestiges of transition. Like a number of others, I personally am not persuaded.

The argument continues. There are caveats to this whole general outline. The archaeological story is not as simple as it may seem: it has been put together by many workers on the basis of study and analysis of prodigious numbers of stone tools at many sites, and it will constantly be revisited by them. I have merely attempted to boil it down, mostly from already boiled-down efforts of authorities. Second, there is no skeletal evidence as complete as the St. Césaire find that can be traced to the very earliest part of the European Upper Paleolithic. The fact remains, however, that the Neanderthals are seen no longer anywhere after 35,000 B.P.

With small exceptions. A colony or so of Neanderthals survived for several thousand more years in southern Spain. A fine, typically Neanderthal jaw was retrieved from the Zafarraya cave near Malaga, dated to 30,000 B.P. This is astonishing, and it brings us to the key question: how are the Neanderthals really related to us?

What Name, Please?

Of course, for the Multiregional Evolution school, they *are* us: *Homo sapiens*, simply the European ancestors of Europeans, in spite of the sharp differences between their skulls and the homogeneous form among all moderns. That view is losing ground.

In formal naming, there had been a general feeling that Neanderthals should be a separate subspecies, *Homo sapiens neanderthalensis*. Now, agreement has been growing that, in the 1850s, William King was right, and that we should recognize a separate species, *Homo neanderthalensis*. This is argued strongly by the highly individual structure of the internal nose, and by the appearance of typical Neanderthal traits in very young skeletons.

Among animals generally, a subspecies, which can interbreed with a sister subspecies, always has the potential to become, over time, a separate species, that cannot interbreed with a sister species if the genetic differences have become great enough. It seems on the face of it unlikely that late human populations might have diverged to such a point — like the separation of horses and donkeys, let us say — in a few hundred thousand years. Nevertheless, many closely related species can interbreed given the chance, like lions and tigers in a zoo, so the criterion is not absolute. Not only is the physical distinctness of the Neanderthals striking. Equally striking is the lack of any sign of interbreeding, or of absorption of Neanderthals by the Upper Paleolithic people, and their survival in spots like St. Césaire or the very late Spanish relics.

So the argument is strong that, after hundreds of thousands of years, divergence of Neanderthals from our ancestors had reached a species level of distinction. We shall see, in the next chapter, similar signs of distinction in the Near East, where both kinds of humanity were present without evidence of interbreeding.

For perspective, we must remember that Neanderthals, like ourselves, were advanced human beings, leaving any *Homo erectus* survivors, like the Solo people, well behind. Underlining this is the discovery of well-made wooden javelins, nicely balanced for throwing, miraculously preserved at Schöningen in Germany. They are dated to 400,000 B.P. and are clearly sophisticated hunting weapons. The makers must have been large-brained Neanderthal ancestors — perhaps like the Swanscombe woman's gentleman friends, or the Reilingen individual — and the spears would do credit to anyone's forebears.

Chapter 13

New Horizons

Here we take the other fork in the road leading away from *Homo heidelbergensis*. It has to do with the rise of moderns in Africa, where Neanderthals never were.

You will remember that East Africa had near-moderns — perhaps a late model *Homo heidelbergensis* — at about 130,000 B.P. These included Omo 1 and Ngaloba. Few though the fossils are, we sense a gradual emergence of moderns in Africa, at a time when Neanderthals were well along in their evolution in the northwest.

The Importance of North Africa

North Africa shows this, whether or not the emergence took place there. We have no knowledge as to how much change was local and how much might have been derived from the south or east, across the Sahara. We can take note that, while only the Mediterranean shore of North Africa is fit for dense habitation today, at various periods in the past wetter climates opened up the Sahara desert to occupation and travel. However, North Africa — usually separated from Subsaharan Africa and bounded by the Mediterranean and the Atlantic — was doubtless an important region in its own right and needs to be viewed as such.

You may recall that the quarry at Ternifine, in Algeria, yielded three lower jaws and a parietal bone dated to the beginning of the Middle Pleistocene, about 700,000 B.P. The pieces are ascribed, with every justice, to *Homo erectus*, as his representative in the area. Further evidence comes from Morocco, mostly in the form of fragmentary finds, largely lower jaws. From a couple of the Moroccan sites (Sidi Abderrahman and Thomas Quarries), almost as old as Ternifine, the pieces recovered have also been assigned to *Homo erectus*. Another jaw, from Rabat, perhaps 300,000 years old, is viewed as a post-erectus archaic. None of this is very instructive; these and other pieces give evidence only of a continuous occupation, whether or not of continuous evolution.

At any rate, we now come to some increasingly significant fossils, recently dated by electron spin resonance (ESR, a new and complex technology). First are important skulls, and some other remains, from the Jebel Irhoud Cave in Morocco. Found here, a child's teeth are of a size greater than those of any Neanderthal or

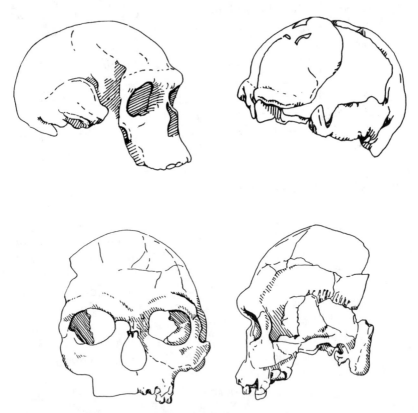

Fig. 51. Northwest African Remains
The Jebel Irhoud remains may be as early as 130,000 B.P. or more. The best preserved, Irhoud 1 (*left*), is somewhat archaic in brow form and lower facial projection, but lacks any resemblance to contemporary Neanderthals in skull or face shape. Irhoud 2 (*right*) lacks a face, but the frontal region shows modernity of form more than Irhoud 1.

The later Dar es-Soltan skull (*bottom*) is also archaic in its marked brows but is otherwise modern in form.

modern; and an isolated arm bone from the cave is similarly massive. The skulls, decidedly archaic in appearance, have been called Neanderthal but show no actual features of that population; they are relatively short, with heavy brows which are, however, modern in form. Narrow nasal bones and a wide face with little projection below the nose are decidedly non-Neanderthal traits. One skull is much like the Florisbad skull of South Africa, which is dated at 259,000 years ago, a surprisingly early age for this specimen which approaches modernity.

The age for the Irhoud people is uncertain but is believed to be 190,000 B.P. This would make the Irhoud fossils distinctly older than Omo 1; and their archaic aspect would conform to such an age. For the present, we must live with these problems of dating.

Somewhat less archaic are a couple of individuals from the Dar es-Soltan cave, also in Morocco. Their age, not very precise, is probably 80,000 years or more. Again, the teeth are more robust than those of any Neanderthal, and the brows are massive. The skull is high, and the face is broad and low. There are no Neanderthal traits, on the word of Jean-Jacques Hublin of the University of Paris, who is an expert on Neanderthal particularities.

Eastward From Eden

If in fact they originated in East Africa, these almost-moderns may have reached North Africa across the Sahara during a moist phase; at least that is a favored opinion. A similar climatic influence might have caused a pulse of similar near-moderns into the Near East, which at that moment would have been like a part of Africa. At any rate, it is here that such near-moderns make their first appearance outside of Africa, from about 100,000 B.P.

It is a long way from Northwest Africa to the Levant (the eastern shores of the Mediterranean), with little evidence lying between. Two poor lower jaw pieces from Haua Fteah in Libya, dated to 47,000 B.P., are archaic but, like the Moroccans, non-Neanderthal — in the only jaw with teeth, there is no retro-molar space. However, Robert Foley and Marta Lahr think they point toward Late Stone Age people of that very region. A good braincase from Singa, on the Blue Nile in the Sudan, long neglected, is now dated at 133,000 B.P., or perhaps the same as Irhoud; and Stringer classes it as pre-modern archaic with a suggestive likeness to the Irhoud specimens. Thus, although the evidence is not very broad, there is

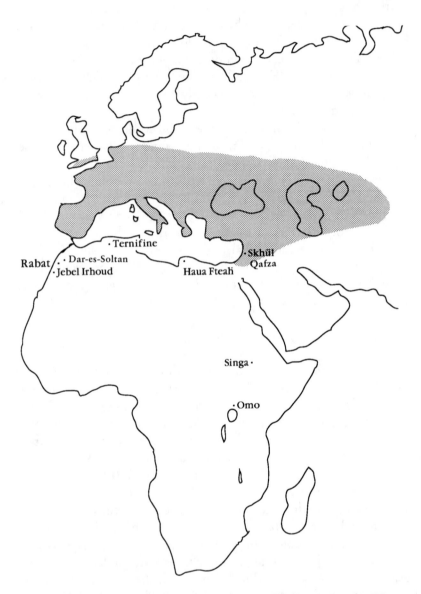

Fig. 52. North and East African Finds in Relation to the Neanderthal Area in the Upper Pleistocene
These are the locations of remains dating from probably 50,000 B.P. to 130,000 B.P. or more, all ascribable to modern (if archaic, in cases) *Homo sapiens*. They are contemporary with, or earlier than, European Neanderthals.

nothing in it just now to puncture the idea of a general North African unity over a long distance into the past, as well as connections, via Florisbad, to *Homo heidelbergensis* in the far south.

In the Levant itself, early evidence is meager. The most ancient specimen, from the Zuttiyeh Cave in Galilee, is 250,000 to 350,000 years old or more; it is a frontal and facial fragment about which there is full disagreement. It is important, because it is poised to be, in various views, an ancestor of Neanderthals, of moderns, of both, or of neither. Given the European evidence of Neanderthal development, above, Zuttiyeh as a Neanderthal ancestor seems unlikely. Milford Wolpoff, of the Multiregional persuasion, sees signs of likeness to the Zhoukoudian *Homo erectus* people, an Asian connection. We might look for comparisons with *Homo heidelbergensis,* as this proposed species has been defined. Zuttiyeh must be put aside but not forgotten.

Qafza and Skhūl: The Jury Comes In

On the other hand, what is clear is that, while proto-classic Neanderthals held sway in Europe, proto-moderns made their appearance here, at two caves near Haifa in Israel: Qafza (dated about 100,000 B.P. and possibly much earlier) and Skhūl (dated about 82,000 B.P.). A good number of skeletons were found at each site, the best such samples comparable to Ngandong or Zhoukoudian. There is no mistaking their general modernity: the bones of the skeletons are straight and slender, and the pelves lack certain clear peculiarities of Neanderthals, while not being completely modern. Likewise the skulls, which are shorter and higher, without marked facial projection. At least one of them (Qafza 9) looks entirely modern; but most have certain archaic marks, like strong brows, large teeth, and so on. These brows are unlike those of Neanderthals, tending, over each eye, to show a division of the lateral part from the center. Nevertheless, in my view and that of most others, the Skhūl-Qafza population as a whole would *not* fit unnoticed into populations of today's skulls — it is *almost* modern.

The character of these people is of first importance. In computerized mathematical treatments (more about this later), one skull, Qafza 6, groups with some late prehistoric moderns, but another, Skhūl 5, is closer to Neanderthals. And Robert Corruccini, a particularly canny user of these methods, finds them all to be *cranially*, but not skeletally, intermediate between moderns and

Neanderthals; he also voices his annoyance with the readiness of most writers to call them simply "modern" and spells out the implications of a more intermediate assignment. In non-mathematical consideration of the specimens, however, it seems to me that the separation from classic Neanderthals is distinct, certainly in the skeleton.

What about the North Africans, at Irhoud and Dar-es-Soltan? They are apparently older than the Skhūl-Qafza people, and in form

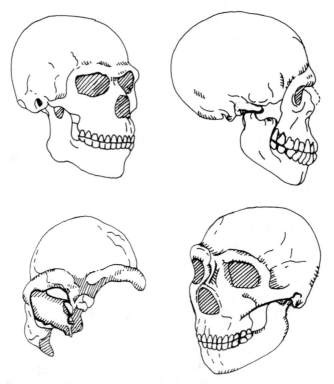

Fig. 53. Modern and Neanderthal in the Near East
Four important specimens from the Levant. Above, Skhūl 5 (*left*) and Qafza 9 (*right*), both dated to 92,000 B.P. or earlier. Skhūl 5 has archaic features, especially the brows (much of the middle face is restored). Qafza 9 is modern in all respects.

Below left, the Zuttiyeh facial fragment, dated to at least 250,000 B.P. Opinion is divided (unevenly) as to whether it foreshadows Neanderthal form. *Below right*, the Amud skull, dated to perhaps 70,000 B.P. It is fully Neanderthal in character, if not as extreme as some others.

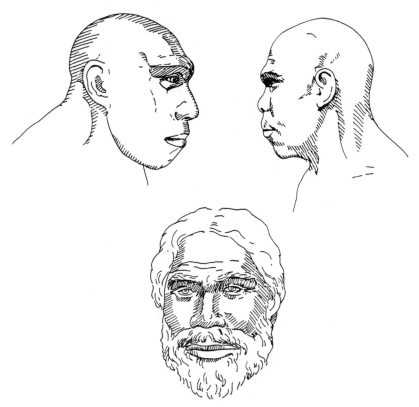

Fig. 54. Some Restorations by Gerasimov
An anatomical (first-stage, of the soft tissues) restoration of a European
Neanderthal, Monte Circeo (*left*), probably a particularly good rendition
of a Neanderthal. First-stage restoration of Skhūl V (*right*). The contrast
with a Neanderthal is marked, in spite of the robustness. Second-stage
("as-in-life") restoration of Skhūl V (*below*).

they seem rather more archaic. However, T. Simmons of Northwestern University, in a metrical, multivariate study, finds the Irhoud
skulls especially similar to some of the Qafza individuals in shape
and not close to any European fossils (certainly as distinct from
Neanderthals as all others mentioned). And results in my own
mathematical studies place Qafza 6 (not the most "modern" of the
Qafza specimens) and Irhoud 1 together, and closer than Skhūl 5
to actual moderns. Finally, Jean-Jacques Hublin has examined all of
the Moroccan fossils closely and perceives nothing in them that

would exclude them from kinship with the Skhūl-Qafza people, nor indeed from an ancestry of moderns generally.[1]

Turnabout

The intruders in the Near East could have come only from Africa. And then, here in the Levant the Skhūl-Qafza people were replaced in turn by typical Neanderthals, some time after 80,000 B.P. A main witness is a skeleton and other skulls from the Amud cave in Israel. Another witness, dated at over 50,000 B.P., is the very good Kebara skeleton, largely complete up through the lower jaw and hyoid bone (Adam's apple) but lacking the rest of the head.[2] This skeleton has the best preserved pelvis of all, showing the elongation of the pubic bone and other traits peculiar to Neanderthals.

A newly found infant skeleton shows clear signs of its Neanderthal nature: again, the presence in very young Neanderthals of such features is a strong argument for their specific distinction from *Homo sapiens*.

Bernard Vandermeersch of Bordeaux, a principal worker at Qafza, believes that Neanderthals made this second appearance in the Near East as the cold of the Last Glacial rendered many parts of Europe inhospitable. At the same time, the moderns disappeared from the Levant. There is no hypothesis to explain this last. There is also no sign in skull or skeleton of actual mixture between moderns and Neanderthals during the thirty thousand-odd years of this second Neanderthal occupation of the Near East.[3]

---

1. For what it is worth, Bernard Vandermeersch, with Olivier Dutour, has found a number of common traits between the Qafza fossils and a 7,000 B.P. series from Mali in the Western Sahara, in addition to similar reflections in the Upper Paleolithic of North Africa.

2. The Kebara find was a burial, as is true of many Neanderthals. But at a suitable time after burial the head was whisked out intentionally, perhaps for ritual purposes. The separation of the skull from the jaw and the rest of the skeleton is too clean to be explained by any other kind of removal.

3. A female Neanderthal skeleton was found in the Tabun cave, next door to Skhūl. It has been thought to have the same or an earlier date. Unfortunately, circumstances of its excavation, and of the deposit itself, prevent determination, and the right date may never be determined unless from the skeleton itself.

Trinkaus recognizes the possibility that Skhūl-Qafza in fact represents a "modern" invasion in the Levant, but in the long occupation of the Shanidar Cave to the east, he sees only signs of the same progression from late pre-Neanderthals to classics as in Europe. He believes that this is unbroken, possibly from as far back as the Zuttiyeh fossil. It may be important that there is no sign of non-Neanderthals having reached the Shanidar sequence before the time when Upper Paleolithic tools appear there, as the harbinger of occupation of the cave by moderns.

But none of this resolves the problem of the arrival of the near-modern Skhūl-Qafza folk at the Mediterranean shore of Israel, and it certainly does not suggest that Neanderthals here were trending toward recent Europeans or other moderns. This has been the crucial disagreement: did Neanderthals and proto-moderns share an ancestor? I think we can be sure of this much: two different populations occupied the Levant during this period, and by implication neither gave rise to the other.[4] Once more, this whole situation, like the later European contact, argues that Neanderthals and modern *Homo sapiens* should be rated as separate species.

But it is a further important fact that, through all this, we find no visible connection at all between differences among people and differences in the tools they made: in the Levant, near-moderns and Neanderthals replaced one another without any clear reflection in tool-making. Instead, there was a wide sharing throughout Europe, the Near East, and Africa of a general Middle Paleolithic tradition, which had a more specifically Mousterian character on all sides of the Mediterranean.

The Deep South

The other side of the coin is the evidence of skulls. We have seen the archaics, showing some signs of continuity and culminating in a modern, Omo 1, at 130,000 years or so. With Skhūl-Qafza, Dar-es-Soltan, and perhaps Jebel Irhoud, he is not alone. But, astonishingly, the other candidates for earliest moderns now known

---

4. Like the earliest writers, Wolpoff thinks all could be accommodated in a single varied population. This is not a widely held view.

anywhere on earth come from southernmost Africa, at Border Cave and the Klasies River Mouth shelters.

In the latter, recent digging has yielded several fragmentary lower jaws, an upper jaw, and a piece of the brow part of the forehead. The cranial parts are all of fully modern form (chins on the jaws, etc.). Most of the fragments are gracile, although a couple are robust, and an ulna (of the forearm) has a Neanderthal degree of robusticity. All were found in Middle Stone Age cultural deposits and are thought to be from 90,000 to 110,000 years old. From similar sources in Die Kelders Cave on the Cape nine isolated teeth are of Neanderthal size but foreshadow modern Africans in form. All this suggests one more example of moderns still exhibiting larger than modern teeth and using a Middle Paleolithic culture distinctly less complex and efficient than its successors.

Border Cave (on the border of Swaziland and KwaZulu) was being dug fifty years ago by a farmer collecting fertilizer. He noticed a skull forepart and partial face, and scrabbled around in his dirt to find more fragments. Given such a casual excavation it is not now possible to relate these bones to the right places in the original deposit; but considerable later work recovered an infant burial and a mandible in better context. These all underlie Middle Stone Age tools of a particular well-known kind, and a number of datings have placed the human material at between 90,000 B.P. and 110,000 B.P., although the latest attempts have reduced the upper limit to 90,000 B.P.

The skull, Border Cave 1, is decidedly of modern form, and in fact has been pronounced to be an ancestral Bushman or other recent African. The tests producing this assignment appear to have been deficient; it is not really possible to allocate the skull to any present people. Rightmire, using multivariate analysis, comes to the same conclusion. Its credentials for antiquity have been challenged repeatedly, with justice. So modern a skull from so remote a time is bound to be challenged.

However, the skull is not a hundred percent modern. The brow, though not large, is projecting as it is in no living Africans and has a telltale feature: it lacks any trace of separation between the medial and lateral parts over the eye, a separation almost always present in recent people. Grave doubts as to the skull's possible age have been voiced repeatedly because of the circumstances of the finding. But this was mostly before the similar age of the Qafza and Skhūl moderns had been established. These were once thought to be much younger, leaving the South African fossils in a seemingly incredible isolation chronologically. Now, such antiquity for moderns or near-moderns seems more plausible.

Thus, southern Africa has a record from *Homo heidelbergensis*, recalling that this species is partly defined by such fossils as Broken Hill and others. It extends into proto-moderns like Florisbad (which has a look-alike in the Irhoud Cave of North Africa — see above) and finally into moderns or almost-moderns at around 100,000 B.P. This record is not easy to read convincingly — or perhaps no one has made enough effort to read it. But it seems to show one more area of advancing moderns, whether by local development or by continuing contact with Africa further north. Certainly, if East Africa is the source of the diverging people of North Africa and South Africa, the contrast between the ultimate descendants, though all are *Homo sapiens,* could hardly be greater.

In sum, all this widely scattered information about the west proposes the following. A single post-erectus species, call it *Homo heidelbergensis*, was present in Africa and Europe, from perhaps 700,000 B.P. In isolated Europe it developed, over several hundred thousand years, into Neanderthals. In East Africa, in much more recent times, it gave rise to *Homo sapiens*. This new species, at some point before 100,000 B.P., was the source of two diverging populations on either side of the Sahara. The northern one, around 100,000 B.P. occupied North Africa and entered the Near East where it in some manner confronted Neanderthals from Europe. The southern one was the origin of present-day Subsaharan Africans. And East Africa probably saw still another offshoot at about this time, moving into South Asia, as we see next.

As a final point, the Neanderthals may be trying to tell us a little more by way of a general parallel. From about 300,000 B.P., we see them evolving in Europe into the large-brained but special classics after 130,000 B.P. We have a generous sample of these last, just what we lack in Africa. There, over the same period, we have racially undefined descendants, also of *Homo heidelbergensis*, among which, according to the authorities, there seem to survive similarities for south (Florisbad) and north (Irhoud) and northeast (Skhūl and Qafza) before moderns set in. This could be a parallel story of forward evolution in the two continents, with "racial" distinctions (not visible in Neanderthals) appearing only late in Africa. African fossils are now too few to tell.

This is all doubtless a much oversimplified scenario, but it will serve as a first framework for discussion.

Chapter 14

Australian Challenge

We have seen almost-moderns in northern Africa, pulsing into the Near East, probably when warm and moist African conditions prevailed there. Under similar conditions, when the Arabian peninsula was not arid, there now seems to have been another exit, out of the Horn of Africa. It resulted in the very early human arrival in Australia — yes, Australia — and by implication in all of southern Asia.

Such an early settlement there has been a new idea in anthropology, and a startling one. In Australia there is now good information — for example, we know that immigrants, evidently modern in physique, had arrived there well before Cro Magnons knocked on the door of Europe. The new findings will shed light on the central question of Multiregional Evolution versus a single African origin of moderns, and probably before we are much older. Read on.

The Evidence

While the focus here is on Australia, a view of the past necessarily also encompasses Southeast Asia, Indonesia, New Guinea, and the islands of Melanesia. Before the time of *Homo erectus*, the islands of Indonesia were emerging from the sea piece by piece. At periods during the Pleistocene, glaciers in the northern hemisphere extracted so much water from the seas as to expose the land mass joining Borneo, Sumatra, Java, and the Asian mainland. This mass is called Sundaland. A separate continent, Sahulland, was formed by New Guinea, Australia, and the island of Tasmania. But the two land masses were never connected: ocean deeps run between them along Wallace's Line. Alfred Wallace, a friend of Darwin's who spent years in the region, saw that, with some exceptions, the land mammals of Asia were confined to the west of this line. Just east of it lies an island zone, known as Wallacea, with a diminished mammal fauna (although two kinds of dwarf elephants once lived there). Finally, in Australia and New Guinea the only existing mammals were marsupials. Though the gap at Wallace's Line is not wide, it was at no time dry land.

Fig. 55. Sunda and Sahul: The Late Pleistocene Far East
In Australasia, western Indonesia was dry land, connected with Southeast Asia as *Sundaland*. Wallace's Line, as now defined (*left dashed line*), following narrow but deep sea channels, separated the eastern islands as *Wallacea*, an island zone between Sundaland and the supercontinent of *Sahulland* (New Guinea, Australia, Tasmania).

Homo erectus apparently never crossed it. In that zone between Wallace's Line and Australia-New Guinea, there are other gaps of open water which, at the best of low-water stages, still maintained widths of 100 miles. Looking at watercraft of recent aboriginals of Australia, it is not easy to see how such crossings could have been made. But made they were, something like 75,000 years ago.

This is interesting and gratifying news, and it confronts us with special problems. A generation ago, when investigators were few and Australian prehistory was in its infancy, it was commonly believed that the aboriginals had not been long in place — 6,000 years might be stretching it. Today, antipodal archaeologists are among the world's best, and the date of first occupation of the continent has been receding rapidly into the past. Solid dates for sites of human settlement in southern Australia and New Guinea go back to 40,000 B.P. or thereabouts, with good indications of higher figures. At fairly early dates people also reached the nearer islands of Melanesia, such as Bougainville of the Solomons.

The earliest signs, at the moment, are to be found in the north. In Arnhem Land, some well-stratified sand deposits have been found, containing plentiful tools. The sand levels can be dated by radiocarbon and by thermoluminescence (TL), a method extremely useful but a little tricky, depending on the accumulation of electrons in sand crystals caused by radioactive minerals. Here in Arnhem Land the radiocarbon and TL dates agree well and are in the right chronological order, proceeding from the top down. The radiocarbon dates give out below the usual limit of ^{14}C, but the TL dates continue down to beyond 100,000 B.P. The stone tools stop in the deposits somewhere between 45,000 and 61,000 B.P., with signs of their presence back at least to 55,000 B.P.[1] Rhys Jones, probably the most experienced worker, has supported a date of 60,000 B.P.

But this pales before a recent report giving a date, of perhaps 75,000 B.P., perhaps earlier, to an astonishing find. This consists not only of buried tools but also of thousands of small cup-like depressions, not much more than an inch across, gouged with some patterning into the face of rocks. These are covered in some cases by

---

1. A small pit containing tools was found, covered by a sand layer having a TL date of 52,000 B.P. Another site places tools at 55,000 B.P.

deposits of soil, which allows the dating, again by TL. They are, thus far, earlier than any such work in Europe, a sensational fact in itself. They have been pronounced to be "art," and even if not pictorial they are certainly ritual objects of some kind.[2]

Second, these dates carry modern Australians well back into the life-span of the Solo *Homo erectus* people, surviving in Java at least until after 53,000 B.P. This is a confrontation like that of moderns and Neanderthals in Europe, but more astonishing, given that the relic is *Homo erectus,* not *Homo neaderthalensis,* a much more advanced species. We will return to this in looking at MRE and ROA again.

Because of the continuity in form of the tools over time, there is every reason to think that the makers were already fully modern people. The tools themselves are not impressive, resembling those in Asia generally. They were for hand use, not for hafting, with one exception. In the north there appeared a form of "waisted blade," with a groove around the middle for binding to a handle and with an edge made by grinding. This is an effective instrument and is the earliest tool showing grinding, and possibly this kind of hafting, for a handheld tool, known anywhere in the world. Otherwise the simple flakes (which must have been supplemented by wooden tools) evidently were adequate for all needs, and in any case they persisted for some 50,000 years.

It has been assumed that, at the time of aboriginal arrival, sea levels were low on the coasts, making water crossings narrower. Possibly the first seafaring moderns, the travelers could have voyaged on bamboo rafts, with weather in some cases driving them, against their wills, far beyond first objectives. The aboriginals made no efficient seacraft in recorded times, but this is not surprising. Once in place, the early colonists would have had no reason to maintain a seafaring tradition.

---

2.   In the same reports, some of the layers associated with stone tools have given dates of between 116,000 and 176,000 years ago, again by TL. The last date is acknowledged to have imperfections. But either case shows how uncertainties could have major effects on interpretations of prehistory. The oldest date would mean that Australia must have been colonized by pre-moderns.

The Physical Australians

So much for stage-setting. A further real enigma is posed by the living Australians and their neighbors. Those of the continent itself may be briefly described as dark-skinned, large-toothed, with straight-to-curly hair. Body build is medium slender, and rather like Europeans in form. In the recent past, however, there were clearly some natives of large size. Above all, there is a slightly archaic construction of the skull, with thick walls, a low vault, and heavy brows. This skull is a central problem.

Elsewhere, the now-gone Tasmanians had generally similar skulls but woolly hair, though this hair is distinguishable in microscopic nature from that of Africans. In other words, dark skin and woolly hair do not necessarily argue special kinship with living Africans. The Melanesians to the east also have such hair, as well as generally Australian-like skulls. Some, as in New Caledonia, strongly resemble Australians in facial appearance; others, especially in New Guinea, are much more gracile.

Thus, there is a certain variety in the whole region which, given its size and long occupation, is expectable. The Australians proper have always stirred debate as to whether they had a single origin or were hybrids of two or three waves of migrants. Some observers have been impressed by local differences, especially as exhibited by the Tasmanians but also by aboriginals of the north of the Australian continent as compared with the south. Other workers behold the variation but are also impressed by the general continuity. In a recent study of skulls, using an intricate computerizing of specific skull details rather than measurements, Colin Pardoe concluded that, among Australian peoples generally, the Tasmanians showed no difference greater than might be expected to result from mutation and chance, given the 8,000 years of isolation since the separation of the island. Analyzing skull measurements in a similar way, I find that, though all three are close, the Tasmanians seem slightly closer to a tribe of Melanesians (the Tolai of New Britain) than to South Australians. This also argues a general early community of origin for all.

On balance, I favor the view of a continuum within one general Australoid branch of humanity, with obvious but natural local differences. Take note that, when seen against other major regional populations, Australians and Melanesians do indeed cohere. The question is not closed but its significance can be overdone.

Here is one special bit of Australian prehistory. After the immensely long archaeological record showing little change from the beginning, a new tradition of small stone tools made its abrupt appearance at about 5,000 B.P. — very late. This introduced smaller, skillfully made flakes, along with the use of these as spear points or knives by efficient hafting to a handle using strong natural gum. This new tradition showed signs of declining in the latest period.

Mysteriously appearing at the same time was the dingo (a wild dog related to the plains wolf of India), the only non-marsupial mammal present in Australia before Captain Cook, other than rats, bats, and humans. But this seeming incursion and the small tool phenomenon have so far turned out to be the key to nothing, historically speaking.

Some Older Australians

Problems are only made worse by past skeletal evidence, which gives us a much wider range of variation than appears in the living. The variation runs along two cross-cutting axes. One is general body size, which ranges from that of today's Australians to something considerably bigger. The other ranges from faces of ordinary size (for Australians) to quite large. Neither axis fits nicely with time.

The shores of what was once Lake Mungo in the former Pleistocene lake system in New South Wales have yielded up two useful specimens dating from 26,000 B.P. to perhaps 30,000 B.P. One was the cremated and smashed skull of a woman, quite delicate in formation. The other, more complete, is that of a man of generally gracile construction, certainly no more robust than modern aboriginals, among whom he would readily fall into place.

Another skull top (WLH 50) from the same Willandra Lakes system seems a little younger. It is very thick and rather low. But the uneven thickness suggests a pathological condition rather than a natural form, and so it may not be reliable evidence. The next oldest is from the Keilor terrace along the Marybyrnong River north of Melbourne. A good specimen, it seems something like a slightly enlarged Tasmanian. There is little remarkable about it other than its age of about 13,000 years.

The Kow Swamp people, at about 10,000 years, bring the surprise. A good number of burials were found at this site. Some of the skulls look like robust examples of today's people. Others —

enough to show that they are no oddity — have skulls of ordinary size adorned by faces looking much too big for them, with really massive jaws. The faces are accompanied by large bony brows and very flat, sloping foreheads. All this is not sufficient to rate these Kow Swamp folk as actual "archaics," or like some of the Skhūl people, but only to give them the prize among recent people for such backwardness. Their facial robusticity simply stands outside other known samples of Australians. There is only one other such case: long ago a stray skull with a similar massive face was found at Cohuna (near the Kow Swamp area) and was realized to have some antiquity. The much later findings at Kow Swamp showed that there had, indeed, been a whole population having this character.

It is important enough to merit a separate paragraph that the *skeletons* of the Kow Swamp people were no more massive than the rather gracile skeletons of today's Australians.

This is not true of male skeletons dated still later, around 6,000 B.P. One was the Lake Nitchie burial north of Sydney. This man had nothing of Kow Swamp about him: he was simply an ordinary Australian on a grand scale, six feet two with a skull to match. The skull of the other man comes from near Cossack on the west coast. This man's head was bigger still, with a very long face which, however, is not of the Kow Swamp type. Otherwise, both individuals are of modern Australian type.

It is this great range of size and form that has kept alive the discussion of different waves of immigrants. Were the first people robust and heavy-faced, followed by some gracile people from elsewhere? Or was there

Prehistoric Australians
Niah Cave, Sarawak 40,000
Wajak, Java ????
Tabon Cave, Palawan, P. Is. 23,000
Lake Mungo 26,000
Willandra Lakes 1500 20,000–25,000
Keilor, Victoria 13,000
Kow Swamp 10,000
Lake Nitchie 6,000
Cossack Cave 6,000

Prehistoric Australians

an intrusion of heavily built or heavy-skulled people into a population as lightly constructed as the modern people? Here the apparent

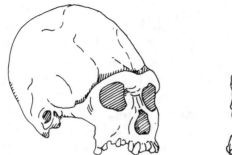 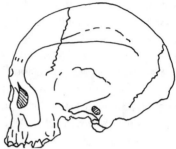

Fig. 56. Pre-modern Crania from Australia
Left, a Kow Swamp individual, showing heavy brows and sloping forehead, but essentially modern and Australoid in character. This is not the most extreme example of the Kow Swamp special character. *Right*, the Keilor skull, 13,000 B.P., with the character of present-day Australians.

new immigration at about 5,000 B.P. could be put into play. But as an explanation it comes up short. The Lake Nitchie man, obviously a personage, was buried wearing a necklace made of a few hundred canine teeth of the so-called Tasmanian Devil; several hundred animals must have died to furnish what was obviously an important ritual object, not a bauble. This nasty little marsupial carnivore became extinct on the continent of Australia, probably from competition with the newly arrived dingo. So the Lake Nitchie man was pre-dingo, i.e., before the hypothetical new immigration of 5,000 B.P. The best date for the Lake Nitchie man is just under 7,000 B.P.

The native Tasmanians were as gracile as the present Australians. Arriving in Tasmania about 30,000 B.P., they were isolated from the mainland after about 8,000 B.P. by the rising waters of Bass Strait. Thus, they argue for an original Australasian population of gracile form. They were affected neither by a Kow Swamp big-faced visitation (before the isolation of Tasmania) nor by any big-headed arrivals around 6,000 B.P.

It may be that the large Lake Nitchie and Cossack individuals were simply exceptional, both in stature and social importance. The Kow Swamp phenomenon is not explained. But all the variation may be a secondary matter. Throughout the past these skeletal recoveries seem to have a basically Australian cast to them, so that, if there were different immigrations, they were probably not from radically different sources. Thus, the question of the number of migrations loses much of its importance. Let us remember that the

Cro Magnons of Europe, before the arrival of farmers, were also robust, in a few cases exceedingly so, compared to living Europeans.

Furthermore, no allowance has been made for possible climatic adaptations. In fact, during the Late Glacial phase of the north, from 45,000 to 16,000 B.P., continental Australia was colder by more than 15 degrees fahrenheit, which is quite a difference. We have seen how body sizes might increase as an adaptation, or at least vary significantly.

Back to Topic A

More than any other region, Australia forces the problem of the two hypotheses: Recent Out of Africa versus Multiregional Evolution. The Australians were the centerpiece of Weidenreich's original formulation of the second view, allowing him to trace a steady development from Javanese *Homo erectus* through the Solo people to the aboriginals, all with their low foreheads and pronounced brows.

Of course, this is an appealing idea. And the devout among the Multiregionalists hold fast to this interpretation, in spite of present evidence that the times are out of joint. Opinion and evidence are now that the resemblances of the Solo and Australian crania are not what they seem.

Some years ago, a keen student of the aboriginal past, the late N.W.G. Macintosh, saw "the mark of ancient Java" stamped on the recent Australians. But after much further detailed work on native crania, he changed his mind. C. B. Stringer of the British Museum is a main champion of the Noah's Ark, or Replacement, school of thought. He argues that coincidence is at work: the common features of the Javanese fossils and the Australians are due to the relative primitiveness of the Australians among moderns and do not betoken any special relationship local to Australasia. Australians are simply especially *early* moderns, that is all.

Another strong voice is that of Marta Mirazón Lahr, at São Paulo University in Brazil. She has exactingly compiled key observations on large numbers of skulls both ancient and modern. This has allowed her to show that the supposed "marks of ancient Java," that is, traits that the Australians are alleged to derive from their special descent from that branch of *Homo erectus*, are just as common or commoner in other living human populations.

However this may be, the aboriginals *are* relatively archaic among modern men and that is the oddity. Nevertheless, they are unmistakable *Homo sapiens*, modern humanity throughout. Their skeletons are modern and, as has sometimes been remarked, their body form is rather like that of Caucasoids. In fact, because of their hair form and general hairiness of body, anthropologists once commonly referred to them as "archaic Whites." They were culturally impoverished, which can be laid to the environment; still, this environment allowed them to subsist nicely on the basis of their hunting techniques and skills. Their hunting skills and tools are impressive: they are the inventors of the boomerang and its beautiful name. Linguistically they are as agile as we are, art is old in the continent, and their social structures and mythology are rich. If, as many now hold, these are qualifications of fully modern *Homo sapiens*, the Australians are qualified.

The important fact remains that Australian skulls are marked by a quality of robustness beyond other living peoples, quite apart from the large individuals like Cossack or the Kow Swamp group. Marta Lahr, with Richard Wright, has made this clear in a further painstaking and intensive study, recording measurements as well as traits of shape like large teeth, brow ridges, depression of the root of the nose, and certain others. These are combined mathematically to produce a few axes of size and robusticity. She finds that, in a general analysis covering these things in a worldwide sample of skulls, the Australians stand apart as specially robust, with their large teeth finding an echo in facial character. This is more marked in large skulls, but holds in small skulls as well. This is putting a thoroughly scientific base under what has previously been left to mere observation and comment. And it will serve us again.

Take the hypothesis, stated already, of exits from Africa by early moderns, beginning about 100,000 years ago. We have the Qafza-Skhūl people, still pre-modern in some aspects or individuals. Then we may hypothesize another population, slightly later, starting from the Horn of Africa, modern but marked by a cranial robusticity probably characteristic of humanity at the time. The same seems to be reflected in the early, but modern, fragments from the Klasies River Mouth caves in South Africa.

This is an appealing fit. Marta Lahr favors the view of "jump dispersals," or movements over significant distances that isolate the migrants from the source, with some evolution or change in the new territory.

Much remains to be found out, especially about the stretch from Africa to Australia. There are only hints of other survivors from such a migration, deeply buried. India, particularly in the south, harbors tribes who stand outside the caste system and its Caucasoid people. Much has been written about them, too much to consider here. Elsewhere there are the Negritos, of small size, in the Malay Peninsula and the Philippines. In the Andaman Islands there are Pygmies who recall the African variety, but who can be distinguished from Africans in various physical ways. By language as well, theirs seems to have been isolated a very long time.

Southeast Asia, from Burma around to Vietnam and Indonesia, has vast populations that might be called "mild Mongoloids," of which more later. There are no solid signs of submerged Australoids (except in some islands closest to Australia), but such have been suggested and would not cause great surprise. After all, in almost any reconstruction of history, Australoid ancestors had to be in Sundaland for a respectable length of time, before and after that Australian date of 75,000 B.P.

Sunda Shelf Life

In broadest terms the Australians give us this. Like the European Caucasoids, they are migrant moderns with a plausible date of arrival. Like the apparent forerunners of Bushmen in southern Africa, they have been preserved in isolation from the time of original entry. And, as things stand, they obviously had immediate ancestors in Indonesia and Southeast Asia, probably before 75,000 B.P.

In reconstructing the past, this is a key area, not only for Australia but also for other parts of Asia. Compared with the well-recorded, fine-grained prehistory of Europe, it is frustrating in the extreme, in spite of considerable work. For one thing, much of former Sundaland and Wallacea, where there must have been plentiful settlement, is now under water. For another, the stone tools are not helpful. As in eastern Asia they are simple in type, with no signs of an emerging Upper Paleolithic, and this has allowed the presumption that the widespread chopping tool, seeming primitive, was therefore early. That presumption has been rebutted by those in the know and, in fact, nothing other than pebble tools can now be associated with *Homo erectus* in Southeast Asia.

There is a little cranial evidence from the Indonesian region. A skull from the Niah Cave in Sarawak on the island of Borneo has been viewed as a gracile Australian in character and is believed to have a date of 40,000 B.P.; neither attribute is entirely certain. Other fragments from Tabon Cave in Palawan in the Philippines, datable to 23,000 B.P., likewise suggest gracile Australians. Also Australian-looking are two skulls from Wajak in Java whose discovery caused Dubois to move there from Sumatra in his hunt for early man. These and other signs demonstrate that such people occupied Indonesia before it was entered by other Asiatics. That is to be expected, and we learn nothing critical from it.

All else is hypothetical. Australoid ancestors had to be in Sundaland for a respectable length of time before, say, 75,000 B.P. At that time dry ground, this had to be the jumping off place for reaching Australia. This would be the critical step: once afloat on any kind of watercraft, progress through Wallacea could have been relatively rapid, as Rhys Jones points out, and Greater Australia (Sahulland), though actually a continent, would have seemed like no more than the next island.

Obviously, Australoids have been succeeded, west of Wallace's Line, by different people, namely the present Indonesians. We do not know how this took place: how much by lineal descent and how much by new immigrants, and this is important to the whole story. If mainly by replacement, this might have been helped along by an enormous volcanic explosion, the greatest ever detected. It happened just about 73,500 B.P. at Toba in Sumatra, and its effect on local populations, from ash in the atmosphere, would have been devastating.

Chapter 15

Looking Backward

Here it will help to change vantage points, looking at the past more through the variety we see in living *Homo sapiens.* This is something we see at first hand, though it is far from giving all the answers. Australia has already given us a quizzical view. Europe, Africa, and Asia each have their special problems of connections, present to past.

Starting with people of the present, we have three kinds of evidence, each of which are deficient in different ways. The kinds are:

External appearance, the "soft parts," showing racial distinctions among existing peoples, which cannot be seen in past populations.

Skeletal evidence, particularly from skulls. This kind can join the present and past people but will always be fragmentary and insufficient.

Molecular evidence, genetic traits of body chemistry and the structure of DNA. The former (of which blood groups are the most familiar) have been devotedly reported on, without revealing much history; the latter, DNA studies, are just being developed and might be the most informative evidence of all. Unfortunately, the possibilities are slight of getting DNA from anyone long dead.

Just now, about the best we can do, apart from studying prehistoric skeletons, is to take inventory of human variation and see what we have.

Race: Information and Misinformation

The faces of race are so obvious that, from antiquity on, everyone could be his own expert. Racism was an unfortunate but natural development, because semantics took over: any people could be called a "race" — not only Whites, Blacks (or Ethiopians), and Yellows (or Orientals), but also Jews, Aryans, non-Aryans — in fact, any human group one wanted to distinguish, for purposes good or evil. Early anthropologists, interested not in politics but in human history, thought that the way to prehistory lay in classifying races in ever more intricate schemes, combining skin color, hair shape, nose form, and so on and giving names to the resulting abstractions. This innocent and misguided endeavor led to a search for individuals best exemplifying such combinations, and thus for the "purest" of this or that race.

No better example could be cited than the long-enduring recognition of Nordics, Alpines, and Mediterraneans as the races of Europe. The Nordic hearthstone was supposed to be Northern Europe, where many people display a combination of blond hair (actually most at home in Russia) and blue eyes (actually most at home in Ireland), tall stature (the Dutch are the tallest), and long heads. The next step was a general supposition that, because most people of an area did *not* fall into such a class, therefore most people were mixtures resulting from formerly pure races in which, supposedly, everyone had once looked alike. And so a lot of speculation went into trying to find the homes and migrations of these imaginary "races." All this rested on a misunderstanding of genetics and of natural biological variation, and the edifice has been dismantled by the very students of some who worked hardest at racial classifications.

But we could throw the baby out with the bathwater. Of course, there are major distinctions in external features among the great regions of the Old World. I will not insult a reader's powers of observation by going into detail, but only point to the following foci. In the Northwest are the "Whites" or Caucasoids,[1] who are very light to brown in skin color, hairy of face and body in males, straight of face, and narrow in the nose. In Africa there are the Negroids (Coon called them "Congoids"), dark of skin, woolly of hair, flattish in nose, and bulbous of forehead, usually with squarish, lobeless ears; they vary greatly in body height. In Australia are the

---

1. So called because the first namer of races, J. F. Blumenbach in 1775, thought that natives of the Caucasus were particularly beautiful; today, this term is useful largely because few people know anything about the Caucasus and so it is innocuous, something precious in these name-conscious days. Similarly, Blumenbach named as races the Ethiopian and the Mongolian, neither of them very appropriate geographically. "Ethiopian" has disappeared, but "Mongoloid" still serves us well, for reasons rather like "Caucasoid."

Australoids, dark of skin, straight to wavy of hair, and hairy of face and body, with heavy brow ridges and retreating foreheads. In East Asia are the Mongoloids, most variable of all. The "classic" Mongoloids — Chinese, Japanese, Koreans — are generally light in skin color but never blond, with straight black hair, little hair on face or body, flattish faces, and eyes largely covered by a fold of the upper lid. To the north of these are people with exaggeration of the same traits; to the east and south are others, many darker in color and with less expression of the Mongoloid features above. So East Asiatics are lumped together as "Mongoloids" but are so varied as to make the term hardly more than relative. The still more varied peoples of the New World are derived, somehow, from the same pool.

Not much can be said to explain racial features as evolutionary adaptations, since little can be demonstrated experimentally. Narrow noses function better to moisten breathed-in cold, dry air, which can be hazardous to the lungs. In high latitudes where the sun is low, lightly pigmented skin lets in more ultraviolet sunlight, which acts to synthesize the vitamin D necessary for bone growth. Hence northern Caucasoids, narrow-nosed and fair, seem adapted to northern climates. Too much ultraviolet sunlight causes vitaminosis and skin cancer; heavy skin pigment thus seems highly adaptive in Africa and Australia. In addition, African woolly head hair makes a good anti-sun insulation. Flat faces and fat-covered eyes are good protection against intense cold, in the supposed northern homeland of the most strongly accentuated Mongoloids, like Eskimos and north Siberians.

These are the high spots of such knowledge, and almost the only spots. They do not say how long any such features took to develop; they give us hypotheses, not history. If all the work with races could have told us human history, we should know that history well by now.

So study of the living does not provide answers but only poses the question I have already set forth. Is this racial divergence ancient or recent? More like a million years or more like a hundred thousand or less? Multiregional Evolution or Recent African? That is to say, was there a separate development from *Homo erectus* on, or instead a recent overrunning of the Old World by rapidly branching stems of moderns moving out of one area? We have to turn to subtler kinds of evidence, the skeletal and the molecular.

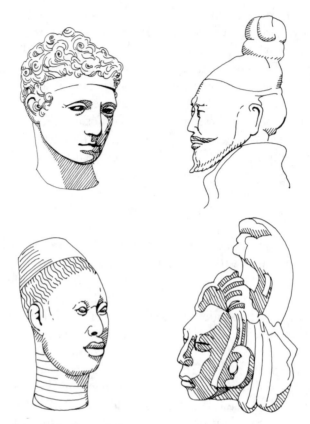

Fig. 57. Portrayals from Major Racial Populations

These heads were executed by excellent indigenous sculptors. Each sculptor evidently idealized his subject but, given the dates in each case, was doubtless not conscious of racial distinctions in humanity outside his own area, and so did not bring any such conception of racial differences to his work.

Upper left, a Roman copy of a Greek head shortly before 400 B.C. This suggests the vertical face, deep-set eyes, and prominent nose of Europeans.

Upper right, a Chinese warrior from the tomb of Shi Huang Ti, date about 0 A.D. The face is broad and flattish, the brows are a straight line over the eyes, which seem prominent because of the low nose between them.

Lower left, a Yoruba portrait head, West Africa, twelfth to fifteenth century. Fullness of nose, lips, and cheeks is evident, as is the full forehead and lack of bony brows.

Lower right, Maya head of a statue from the Hieroglyphic Stairway at Copan, Honduras. American Indians varied greatly, but this head reflects a retreating forehead and a prominent upper nose, generally typical in the Americas.

One general question can be asked of skulls. Living races have seemed very different to many writers. A couple of these, early in this century, thought the main "races" were different species, descended separately from orangs, chimpanzees, and gorillas. Analysis of skulls will easily disabuse us of such a foolish idea. Although some anthropologists are good at this, I myself have trouble recognizing the probable origin of a given skull by looking at it. But you can get an objective grasp on skull shape by taking a sizable number of measurements and falling back on a computer. Combining the measurements from many groups of skulls, the computer can rearrange the information to produce new "measurements" or scores on a few new axes or scales, revealing the essential kinds of variation that lie behind the measurements you have chosen to take. This is known as multivariate analysis. The operation can point out the most important kinds of differences between different populations, say Chinese and Japanese, or even between subgroups within these.

This is not the sharpest tool imaginable, but it provides arithmetical "distances" between known groups. When it is applied to good sets of skulls drawn from all over the world it suggests that, under the surface, living people are much more alike than they seem on the outside. When introduced as a control, Neanderthals fall wildly out of modern bounds and do not point to any present people as more likely to be Neanderthal-related than any other. In the same way, different world regions are roughly equal in their skull distances from one another, though it does seem as though Mongoloids and Australoids are the most mutually unlike.

Europeans: The Upper Paleolithic Phenomenon

Embedded in these matters is the problem of the Upper Paleolithic, a cultural horizon of transcendent importance. It was simply taken for granted in an earlier chapter, as marking the arrival in western Europe of migrants who replaced the Neanderthals. In fact, the purview of early archaeologists was limited to Europe; they were dealing with the end of the story, not its beginning.

Those earlier archaeologists took due notice of the more refined and complex stonework, the bone tools and the art objects, as the Aurignacian culture of the Upper Paleolithic succeeded the Mousterian culture of the Neanderthals. Lately the scholars have been re-emphasizing the contrast and its implications for behavior.

They recognize not only the new richness in Aurignacian deposits, both utilitarian and artistic, but also the signs of greater interest in new materials. Well-worked bone tools appear. Red ocher was used lavishly, and objects of personal adornment were abundant: sea shells, ivory beads, animal teeth, and carved bone appearing in necklaces or pendants. These were followed later in the Upper Paleolithic by the great cave paintings. The Neanderthals used red ocher, perhaps for painting their skins, but the other materials are foreign to Mousterian remains. Sea shells, and even stone materials for tool-making, betray an Upper Paleolithic trade extending over considerable distances. There was greater specialization in animals hunted, especially reindeer. Living sites give evidence of much larger settlements and, thus, of heavier population. The sites also manifest more organization and better housekeeping; the Mousterians, apparently, were less fastidious, throwing only the largest meat bones right out of the camp.

So impressed by the contrasts are the archaeologists that they talk about a manifest surge in human capacity, a sort of behavioral reorganization, as an actual step in our most recent evolution. One explanation in particular is voiced: an evolutionary advance in linguistic ability, through a new flexibility in syntax, and thus a greater competence in the handling of complex concepts. (This has been discussed earlier, in the question of early language and *Homo habilis*.) The suggestion is that this last step marks the emergence of fully modern *Homo sapiens*, and is frequently called the "sapiens explosion."[2]

The phenomenon has been mulled over by some of our best mullers. What does it mean biologically and culturally? Biological change necessary for such a shift could hardly have been as sudden as the appearance of Upper Paleolithic culture. It must have happened earlier, and supposedly before any dispersal of modern *Homo sapiens*. Indeed, we can see this advanced level of behavior, expressed

---

2. Trinkaus adumbrated the general idea in his 1983 book on the Shanidar Neanderthals, though more in anatomical terms. He concluded: "These interpretations of changing behavior patterns derived from shifts in human anatomy during the latter part of the Pleistocene in Europe and the Near East suggest that Neanderthals were the most recent participants in a level of cultural adaptation that was significantly less complex and efficient than that of anatomically modern humans."

in art though not in the always simple stonework, coming to Australia with the first aboriginals, much earlier than the arrival of the Cro Magnons in Europe. In addition, good bone points, including barbed harpoon heads, have been excavated along the Semliki River in central Africa; they have dates of 90,000 B.P. or earlier. So let us be warned that the Upper Paleolithic Europeans have no copyright on the apparent "sapiens explosion." It would appear instead that the European Upper Paleolithic was more a ripening expression of the new human level, not a cause of it, and we need much more evidence on the nature of this florescence.

Whatever the hypothesis, the Upper Paleolithic is change in the concrete. The hearth of the transition is not known precisely but the archaeologists are closing in. As Harvard's Ofer Bar-Yosef has emphasized, the Near East long manifested only the same Middle Paleolithic culture, Mousterian in general character but locally varied, in the hands of both the Skhūl-Qafza people and the local Neanderthals, with no significant differences in the tool forms. Then and later, the Mousterian of the Levant and North Africa was also similar. The whole area, including the Upper Nile, was the eventual seedbed from which Upper Paleolithic stonework emerged, although specifics are not yet evident. Desmond Clark (at Berkeley) sees a likely direct evolution in the Negev, on Egypt's eastern border, as the Ahmarian culture, so-called, developed right out of the Levallois point of the Mousterian. But transitions are not really clear. Bar-Yosef, for example, points out that the earlier Neanderthal site of Kebara shows that the settlement was not as benighted as a simple Middle-Upper Paleolithic contrast would suggest: the site shows organization as to hearths, etc., and the stone industry contained blades. In the Levant, in fact, such Upper Paleolithic ideas in stonework were expressed at several points in the Middle Paleolithic, though without continuous development.

Paul Mellars of Cambridge University thinks the transition is earliest and clearest in the Levant. At Boker Tachtit in the south, the change is apparent beginning in levels dated to about 47,000 B.P., and at Ksar Akil in Lebanon the techniques of flake production are seen shifting to the new at about the same date. Ksar Akil has also provided, though not from the lowest levels of all, the skull of a completely modern child, which may be the earliest such association with the Upper Paleolithic now known. But this does not pinpoint the origin of the widespread Aurignacian, which appeared before 43,000 B.P. in Eastern Europe, apparently with real anatomical moderns. Upper Paleolithic cultures also emerged in due course in North Africa, though having a different character.

This is a juncture pertaining to the northwestern quarter of the Old World, the region we have been considering. But, again, it would be a mistake to equate Upper Paleolithic emergence in that region alone with the "sapiens explosion." In fact, a similar advance to the Later Stone Age took place in southern Africa, apparently at least as early. But we must finish with Europe.

The New Europeans

The new people of the Upper Paleolithic are not only modern, they are recognizably European. They are not as "modern" as living Europeans: they have bigger teeth and more robust skulls. But these skulls simply look European and, when tested by certain complex analyses of measurements, yield one of two results: either they ally themselves with living Europeans, as we shall see presently, or their robustness removes them from alliance with any present population.

These are the people we have been calling Cro Magnons, after that famous French site, although the name now stands for many Upper Paleolithic skeletons from all over Europe. They round out this history of the northwest of the Old World. But we do not at the moment know where they first arose; we may assume, on the basis of some very unsatisfactory bone fragments, that they were present in eastern Europe with the earlier Aurignacian there. No place to the north or east looks promising as a source of the Cro Magnons. The Ukrainian Mousterian population, surely Neanderthal as judged from a couple of remains, was succeeded by an Upper Paleolithic culture that was not Aurignacian but was not necessarily locally evolved, i.e., probably an immigrant from elsewhere. Otherwise, the colder parts of Eastern Europe and Russia were populated poorly if at all before the Upper Paleolithic. In fact we do not know with any degree of certainty where most of the present people of the Old World came from.

On a lumpier scale, we might notice that, as soon as glacial retreat permitted, other aspects of culture advanced rapidly, to agriculture and urban life, in the Near East, East Asia, and the Americas. Nothing like this happened in the ameliorated conditions of the previous interglacial before 75,000 B.P., whether among Neanderthals or early moderns. So if the archaeologists want to

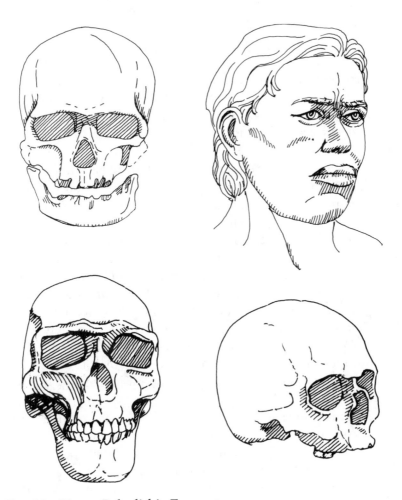

Fig. 58. Upper Paleolithic Europeans

Above, the first Cro Magnon skull, found in 1868 at Les Eyzies in the Dordogne. Though robust, it shows characters of later modern Europeans. The skull is high, the face is vertical, and nose and chin are prominent. The restoration of the living from this skull is by Gerasimov.

Below, early Upper Paleolithic East Europeans (from Předmostí and Mladeč, Czechoslovakia). Some believe these skulls show Neanderthal features, suggesting that they are hybrids, or transitional from Neanderthals to moderns, demonstrating continuity. Other workers find such traits instead to reflect simple robustness, since individual features are specifically modern and since mathematical testing allies the specimens with modern Europeans.

argue a new surge in human capacity at some point during Last Glacial times, who are the biologists to say them nay?

The People of Europe

Cro Magnons arrived in western Europe about 35,000 B.P., replacing Neanderthals. After that date, no non-modern "archaics" are known from any part of the world. In fact, this date is rather late for moderns generally, and so their origins must be looked for outside of Europe.

Furthermore, these origins obviously have to be older. If Upper Paleolithic people were "European" from about 35,000 B.P., then such major population distinctions are at least that old. And the Cro Magnons were already racially European, i.e., Caucasoid. This has always been accepted because of the general appearance of the skulls: straight faces, narrow noses, and so forth. It is also possible to test this arithmetically. Using the methods of multivariate analysis described above, we can take a single skull and compute its generalized distance (a sort of sum of the best information about its shape) from known world populations of the present. This can say which living population the skull is least distant from and will also state the probability of its belonging to that population or to others. The method is not absolute, and makes some misjudgments, but it is strong information all the same.

Not many ancient skulls are in good enough condition for all the measurements one needs for the best comparison. Five Upper Paleolithic specimens are available to me for study: Mladeč 1 and Předmostí 3 and 4, all from the former Czechoslovakia, and Chancelade and Abri Pataud from France's Dordogne. Except for Předmostí 4, which is distant from every present and past population, all of these skulls show themselves to be closer to "Europeans" than to other peoples — Mladeč and Abri Pataud comfortably so, the other two much more remotely.

The evidence from later European crania is equivocal. These date roughly from the end of the Pleistocene, about 10,000 years ago. A group from Muge in Portugal readily identifies itself as European. Other important specimens, from Brittany and North Africa, do not so identify themselves through multivariate analysis, although from general appearance they have always been rated as "Caucasoid." Again, robustness probably plays a part.

A different sort of overview results from the work of Ilse Schwidetzky and her associates at Mainz University. They computed multivariate distances using very large numbers of European skulls but taking fewer measurements on each, covering evidence from the Mesolithic (the last pre-agricultural stage of culture) to Roman times, a span of 10,000 years. Unlike the impressionistic constructions of earlier times, Schwidetzky's analyses showed no great heterogeneity among Europeans but rather a gradual shift moving from south to north, from broader to narrower faces, as if Cro Magnons had given way to somewhat more lightly structured Europeans.

These findings are a far cry from the somewhat Wagnerian scenarios of distinct European races transporting distinct cultures, gods, and world views. Caucasoid origins remain, shall we say, shrouded in mystery. But as far back as we can see continuously, Caucasoids have occupied post-Neanderthal Europe, North Africa, the Near East,[3] and much of India.[4] We do get some light on how the pot has been stirred in later prehistory, which helps the process of demystification. The information comes from very interesting studies involving the troika of language, agriculture, and genes.

ABO
MNSs
CDE (Rh)
Kell
Lutheran
Duffy
Hp
Tf
Gc
HLA
AK
ACP
Gm
PGM

Designations of Well-Known Blood and Enzyme Gene Systems

Take genes first, these being the ones controlling blood and protein types. Almost anyone can give you his ABO group and his main Rh type (Rh+ or Rh-). Many other such genetic variants can now be tested for, though few of them are like the ABO groups in posing risks in blood transfusions. Information has been collected in great amounts, and populations are compared as to the proportions

---

3. It is an open question whether the Skhūl and Qafza people foreshadow Caucasoids; I am inclined to think they do, but that is an opinion, not evidence.

4. The great majority of the Indian population is Caucasoid in facial and cranial form, and the dark skin, hair, and eyes must be ascribed to adaptation to intense sunlight plus gene assimilation from earlier inhabitants — a rather obscure situation.

they exhibit of the genes in any one gene system (e.g., the percentage of A or B or O).

A population's gene proportions change over time, for different reasons: chance, but also by adaptation. As an example of the latter, some gene types evidently protect against malaria and are thus under selection, but most such disease connections are not understood. Some gene systems are evidently rather stable over longer periods and so may be the ones best reflecting situations in the past. At best, however, in genetic terms "the past" means fairly recent history. Therefore, genes have not figured much in discussions of earlier *Homo sapiens*.

Furthermore, the data are difficult to use. Multivariate distances may be computed, like those using skull measurements. But because of the different degrees of stability from one gene system to the next, results are affected by the choice of systems. In any case, please hold this bridle while we round up language and agriculture.

At some much-debated date the Indo-European languages spread into India and all through Europe, in which continent they replaced every earlier tongue except Basque. After 200 years of study, the point of origin of the Indo-European stock has still not been agreed on, and just now there is ardent dispute as to whether the spreaders were recent horse-riders coming out of the Russian steppe north of the Black Sea at about 6500 B.P., or early farmers entering Greece from Turkey at about 9,000 B.P. — quite a discrepancy. (Indo-European languages themselves have been replaced by later immigrants speaking non-Indo-European languages in Turkey, Hungary, and Finland.)

Robert Sokal and his associates at Stony Brook State University (New York) have been ingeniously measuring linguistic and genetic "distances." They find that people belonging to different European language groups of today show greater distinctions in their total gene proportions than could be explained by simple geographic distance and accompanying drift in genes. That is to say, the present genetic picture is still reflecting some differences in origin among the speakers of different language groups. Sokal and his group conclude that: 1) actual migration was important in the spread of European languages, that is, they were not just passed along by a kind of missionary influence or else imposed through conquest by a small group; 2) the effect of this early migration still persists, and 3) some genetic amalgamation occurred between the migrants and earlier people.

Lately, Sokal and his associates have taken information from the archaeologists about the spread of agriculture into Europe from the Near East, information which by now is pretty full. Plotting this information and superimposing it by mathematical methods on their other plots, they find that it specifically supports their earlier conclusions about the pattern of movement of languages and genes. But on the basis of these analyses, the Stony Brook group remains unhappy with either of the prevailing hypotheses as to Indo-European origins and dates of migration.

Still, without giving answers in detail, all this thickens up an interpretation of movements of generally related peoples into Europe, who introduced agriculture to those already there, the immigrants also establishing themselves through increase in their own numbers. This is simply a useful view of the growth of European peoples through migration and cultural reinforcement. Only by inference could this model be applied to the earlier expansion of the Cro Magnons at the expense of Neanderthals or, in other regions, to replacement of *Homo erectus* by *Homo sapiens*, the winners in each case getting a boost from cultural advantage.

North Africa today is also "European," that is, Caucasoid, but less is known about the foundations of this. From what we saw of pre-moderns there, North Africa might in fact be the Caucasoid source and not, like western Europe, the end of the road; this is a question for the future. In any case, North Africa eventually received an entirely different language family, the Afro-Asiatic, which includes Egyptian and the Hamitic and Semitic language groups, the last including Arabic and Hebrew.

Chapter 16

Out of Africa?

And Adam called his wife's name Eve; because she was the mother of all living. (Genesis 3:20)

A frica is the well out of which all primates seem to have flowed. We have seen human history there, to the threshold of fully modern people, in East, North, and South Africa. There are important later fossils, but let us look at them after bathing in a special controversy over Mitochondrial Eve, the mother of us all. This stems principally from a now-celebrated study of 1987 by Rebecca Cann, Mark Stoneking, and Allan Wilson, all then at Berkeley, and calls herewith for a small refresher of one's high school biology.

Firstly, every living cell consists of a cell wall, cytoplasm (Greek for "cell stuff"), and a nucleus. In the nucleus are the chromosomes (twenty-three pairs in ourselves) which are simply divisions of the total length of the DNA (DeoxyriboNucleic Acid) chains. These paired chains (the "Double Helix") are made up of sequences of chemical bases (only four kinds of these are involved). The full length is about three billion such base pairs. The genes, controlling all development, are strings within this, to the number of about 100,000 of such strings. Accordingly, the current proposal to read the whole length of human DNA is an effort almost comparable to going to Mars.

We, as living things, grow by cell division. Before a cell divides, the chromosomes in the nucleus line up by pairs and all divide into exact duplicates; then the nucleus divides, each half taking a full set of chromosomes, and the whole cell follows suit. Any error in the chromosome duplication is a mutation, with varied, or no, effects on the resulting organism.

Secondly, outside the nucleus, in the cytoplasm, there are mitochondria, something under 2,000 of them to a cell. These are little engines changing food into energy. Each mitochondrion has a minute bit of DNA (mtDNA), distinct from that in the nucleus of the cell, this bit being identical in all the mitochondria of any individual. The number of base pairs in this mtDNA is about 16,500, not three billion as in the chromosomes. This number is so much more manageable that its analysis has attracted a lot of recent interest.

Thirdly, in reproduction the maturing egg and sperm each get only a half set of their parental chromosomes, so that the fertilized egg, the new individual, has a new combined set with half from each parent. This maternal cell, however, also constitutes the first cell of the new embryo, which begins to divide. The sperm cell is essentially only a swimming nucleus, and such mitochondria as it may carry do not take any further part. Thus, whether to son or daughter, the mitochondria come only from the mother in every generation. Hence the allusion to Eve.

But why the Mother of Us All? This is an inadvertent confusion. Here we come to the Berkeley studies. Cann et al. used special enzymes to cut the mtDNA into shorter fragments. They analyzed the varying patterns of fragments in different individuals, finding a large number of separate types. In their very explicit paper, they grouped the types in various ways, but all of these led to the conclusions that 1) in Africa there was greater diversity than elsewhere, with more types not found outside Africa, 2) their best "tree" of types had two primary branches, one being entirely African and the other containing all other types including some Africans, and 3) that the pattern seemed to show, in other parts of the world, branchings such as would occur locally at times after the major African-other separation had taken place.

New types appear by mutation. But also there is an opposite effect, a constant loss of types because a type is handed down only from mother to daughter, and some mothers have only sons, who cannot pass the type along. And so, in a small group, this effect can go on until only one type survives, although this is countered by fresh mutations. As an analogy, you might think of long-isolated small European communities in which some paternal family names "daughter out" and are lost, in families with no sons, while other names become almost universal; if this should persist long enough everyone would in fact have the same surname. In the analogy, the sexes are reversed: with mtDNA it is the mother who has no daughters, but only sons whose mtDNA type is therefore lost.

So the Mother of Us All does not mean she was the only female ancestor; everyone has to have many female ancestors. It simply means tracing back to a single mtDNA type. We are all descended from some woman who had that type, but also of course we descend from her mother-in-law — who could not transmit her own type through her son — and similarly from many other women.

In this sense Cann et al. interpreted all later variation as having arisen from one postulated type or "common ancestral female." Instantly and predictably, this was translated by others into the Mitochondrial "Eve" — in fact, in the same issue of *Nature* carrying the 1987 article, and of course throughout all print since then, although the Berkeley people disowned the idea.

The inference of an African origin was promptly contested. However, it was supported in a later study by Linda Vigilant and others, also under the leadership of Allan Wilson.[1] Published in 1991, it overcomes some of the limitations of the first work. The new study worked directly on a small segment of the mtDNA itself — 610 base pairs — rather than on fragmentation patterns; it included more diverse Africans; and it introduced a chimpanzee. This ape's DNA sequence is very far removed from any of the human types, but is closer than that found in any other non-human form. The 1991 analysis pointed to the only likely overall tree as one rooted in Africa.

As in the first study, the basic data were subjected to a computer analysis to produce trees of shortest distances among types. Figure 59, reproduced from the report by Vigilant et al., is one of a number of closely similar results reached by them. It shows a clustering of 135 types. There are actually 189 *individuals* included, but a number of them share *types* (that is, the end boxes in the diagram may include more than one individual). It is striking that such sharing appeared only *within* populations, which strongly implies that such types are the result of more recent mutations occurring in populations already separated from others.

This clustering has twenty-five nodes, or junction points, not six as in the first study. Taking possibly impermissible liberties, I would group these arbitrarily so as to get the most striking result. Here is my compilation, consolidating all into six superclusters, and going from least inclusive to most inclusive:

---

1. Of the original group, Allan Wilson, the prime mover, has died; Linda Vigilant, Rebecca Cann, and Mark Stoneking have all gone to other universities. Vigilant's 1991 co-authors were Stoneking, Henry Harpending, K. Hawkes, and Wilson.

		Total
Last cluster	7 New Guinea, 1 Asia	8
Next 9 clusters	4 New Guinea, 7 Asia, 8 Europe, 1 Africa	20
Next 4 clusters	5 New Guinea, 13 Asia, 7 Europe, 9 Africa	34
Next 1 cluster	1 Asia, 22 Africa (including 9 Pygmies)	23
Next 3 clusters	1 New Guinea, 1 Australia, 2 Asia, 24 Africa (including 12 pygmies)	28
Next 8 clusters	22 Africa (including 15 Bushmen, 6 Pygmies)	22

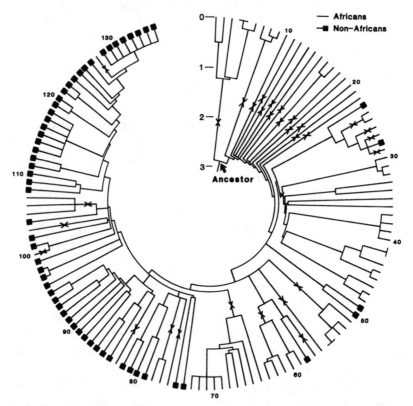

Fig. 59. The "Tree" from Mitochondrial DNA for 189 Individuals
The 189 individuals yielded 135 different mtDNA types. The length of the
stems reflects the degree of dissimilarity among the individuals. Types are
shown only as "non-African" (black squares) or "African" (plain lines).
"Africans" included some black Americans. Thus, though seemingly spread
out toward the end of the line, New Guineans there are actually fairly closely
related, while Africans, starting at 12 o'clock, are most unlike one another
and most different from all other populations. Non-Africans, of any kind,
are furthest removed from the root, and from African lineages in general.

This clustering emphasizes, much more than that of 1987, the separation of Africans in the primary cleavages. And a similar later examination of the same materials, by Roy D'Andrade and Phillip Morin, produced an even more extreme separation of Africans from others than mine shown above.

Not So Fast, Please

I cite the Berkeley studies in some detail, not because they are the last word but as the entry into this important field and its controversies. Their scenario of Africa as the Garden of Eden from which modern humanity went forth to replenish the earth is a happy one for the Replacement school. But it is less than Scripture to others, especially those of the Multiregional Evolution persuasion. Serious objections have been made on the grounds of method and of timing.

The first objection is statistical in nature. With well over a hundred individuals tested, the number of possible "trees" like the one in Figure 59 — that is, the number that would be produced if the distance from each individual to every other were tested — is almost infinite. Drawing up a single one, as here, inevitably creates a totally false impression of precision. What Vigilant et al. have done is to use a computer program to pull out those trees in which the sum of the distances among individuals is the smallest; here, for example, the New Guineans do indeed draw close together, which is persuasive. Thus they selected a likely case among many; they had to come down somewhere.

Critics raise questions of technical assumptions, in methods of analysis, not dealt with in the study. For example, Maddison, Ruvolo, and Swafford, a mainly Harvard group, were given the basic data by Vigilant et al. Reworking these, they found many equally short "trees," some of which had an African root and some of which did not, in fact being rooted in New Guinea! This seems to stand the original result on its head but, as may be seen, the list of individuals included was rich in African pygmies and Bushmen and in New Guinea natives, so that the nature of these populations may affect the outcome. The theoretical points of analysis are complex, and a lot of hard work remains to be done. As Maddison et al. and others demonstrate, an African center of origin is not disproved but also is not nailed down. Be that as it may, continuing studies reinforce the suggestion of African separation and primacy.

Timing is the second problem. Estimates of time depend on rates of change, i.e., mutation. The authors of the original study (Cann et al.) assumed mutation rates of 2 to 4 percent per million years, judging from mammals generally but also from differences among New Guineans in their sample, who were taken to have first moved from Asia to New Guinea at least 30,000 years ago (as we have seen, it was actually more). These rates gave them estimated dates between 140,000 and 290,000 years for "Eve," the common lineage — the Big Bang of this scenario — and a time of leaving Africa of 90,000 to 180,000 B.P. for the first moderns to do so. Evidently, "Eve" herself need not have been as modern as the eventual emigrants. Other evaluations since then have produced similarly wide ranges in dates: Vigilant at al., with an assist from the chimpanzee (meaning rough estimates for the time of chimp-human divergence), suggest a basic modern ancestor (Eve, if we must) at between 166,000 and 249,000 years ago, before an exit, at some later date, of the ancestors of today's non-Africans.

These figures have been strongly contested. Opponents think the assumed mutation rates are too high: if they are lower, the times are longer, and so Eve and the leaving of Africa would be pushed back into the time of *Homo erectus*. This, obviously, would suit the book of the Multiregional Evolution people: it would amount to one early episode of migration out of Africa, rather than two episodes, one early and one late. Cann et al., however, insisted that the pattern fits the Out of Africa model best, with replacement of *Homo erectus;* any mixture of the migrating moderns with such earlier people would have introduced highly divergent mtDNA types into Asiatics and Australians. Signs of this have not been found.

A New Free-For-All

These discussions are based on what we know from living peoples, not on fossils. Controversy has been expanding along similar new lines: other kinds of molecular evidence, combined with theoretical-mathematical considerations.

Quite recently, all this has become a small industry in itself, of some complexity. Briefly, viewing diversity in several different genetic markers has led workers to estimate dates of divergence among non-African populations, estimates which converge on about 140,000 years ago. A few have higher possible limits, to

300,000 years or more, but none are as high as a million years, such as the Multiregional view would demand. That is, dispersal from Africa at the time of *Homo erectus*, and subsequent local evolution, would demand greater amounts of mutation and diversity than are found.

Other considerations are more complex. The MRE model of broad evolving populations throughout the Old World is alleged to demand unrealistically large numbers of people. Henry Harpending et al. have detailed at length the factors of different population sizes, and of expected bottleneck effects, on patterns of mtDNA seen today. Also, the apparent likeness of Australians and Africans in cranial measurements needs rescaling in the light of relative population sizes. They feel that no model clearly prevails as yet, but that a modified Out-of-Africa model, with special later histories (as with Australia, above) fits best. And Marta Lahr suggests general patterns of migrations of populations followed by separations and some extinctions.

This is all new and rather complicated to detail here. It is probably only the beginning of a good deal of such high-level analysis in the immediate future. Remonstrances from the MRE bench will be heard. And in fact the fossils will be needed as well, unavoidably. Without them, estimates of time remain estimates.

Early African Moderns

The other side of the coin is the evidence of skulls. In South Africa we have seen virtual moderns at 100,000 B.P. or more, in the Klasies River Mouth and Border Cave remains. There is nothing to refute a long-continuing subsequent occupation there by the same population. Archaeologically, the Later Stone Age (after 35,000 B.P., and coeval with the European Upper Paleolithic with which it shares general nature) suggests that the Khoisan people (Bush-Hottentot) then occupied all of southern Africa. A few skulls from this time span, from the Cape to Zimbabwe, can all be identified (by multivariate analysis) as "Bushman."

This striking evidence of slightly robust but certifiable moderns having considerable antiquity hardly means that here is the home of fully modern *Homo sapiens*. It does, however, suggest an early colonization of the area from Eve's Place, *after* modernity had arrived in the species, followed thereafter by isolation. H. J. Deacon of Stellenbosch University emphasizes this isolation, noting from archaeology that southern Africa remained a region of stone age hunters until the nineteenth century arrival of Europeans and of Bantu cattle-drivers, bringing today's history. The Bushmen, except in the Kalahari Desert, have all but disappeared, sunk into the "Cape Coloured" population of the south. But until the last century, the area apparently served as a refuge for one segment of the earliest moderns, a refuge approached in isolation only by Australia.

It is elsewhere in Sub-Saharan Africa that we would like to know more. There is almost no evidence for western and central Africa. However, some fragmentary pieces of a number of individuals were found at Ishango, on the Semliki River at the border of Zaire and Uganda, very much at Africa's center. They are dated at over 25,000 B.P., and appear to represent robust Negroids, that is, like the present people of the region. Such an identification would suggest that this major population became established at least by the later Pleistocene. It would also indicate that a distinction already existed between a Negroid and a Bushman people.

Furthermore, at much earlier sites along the same Semliki River, dated to at least 90,000 B.P., knives and well-formed barbed harpoon points, made of bone, have lately been found by the team of John Yellen and Alison Brooks. These otherwise appear in the far later European Upper Paleolithic (see the preceding chapter), as one of the supposed marks of the "sapiens explosion." Thus, we have

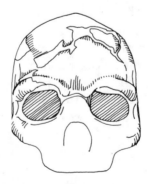

Fig. 60. The Border Cave Skull
These fragments of skull and part of the face have been dated to both sides of 100,000 B.P. Though not clearly assignable to any modern population, the skull cannot be excluded from ancestry of later peoples of South Africa.

possible hints of early moderns here in central Africa, poorly known and not keyed in to other regions. More finds are urgently needed.

One aspect of African history we have in hand: the spread of the Bantu-speaking peoples in the latest millennia. Their languages are all closely related and are traceable to a parent among the Niger-Congo family of languages in West Africa. With the arrival in Africa of agriculture and cattle, Bantu-speakers expanded through much of East Africa and all of central and southern Africa.

But we know far too little of events before that. We must assume that, before any Out-of-Africa exodus, some local differentiation was already taking place in Africa, leading on the one hand to peoples more like Africans as we know them, and on the other to peoples less like them. Where would the latter be? Northwest Africa may have been the source of those almost-modern Levantines, Skhūl and Qafza, but not necessarily the scene of their actual evolution and appearance. Is East Africa still the important focus? Older analysts of "race" constantly noted a less "African" appearance of people, however dark, reaching from the Horn of Africa northward and were given to talk of a "Hamitic strain," that is, admixture from Caucasoids from the north. Perhaps these scholars had it backwards.

Good cranial evidence is lacking. Long ago, Louis Leakey excavated some caves in Kenya, finding a few skulls of no great age, this being reckoned at about 7,000 years. He and Coon, both good at eyeball estimates, thought the skulls were most like late Stone Age crania of North Africa, especially those associated with the Caspian culture of Afalou-bou-Rummel. Marta Lahr has studied this last group in detail, and finds it to be fully modern but stamped by the robustness that marks various late prehistoric peoples. This phenomenon needs fuller examination.

I have subjected some of these skulls to the kind of mathematical analysis I have described. They are refractory, not revealing apparent connections with any particular living people, African or other; this also is true of those Afalou North African crania. I have done similar analyses of more recent crania from the same region, which identify themselves primarily with modern Bantu-speakers of Kenya. This is wispy evidence but could be read as suggesting the replacement, in quite recent times, of an earlier, less "African" population by the expanding Bantu. In other words, Sub-Saharan Africa seems to have been increasingly Africanized, so to speak, in the last few millennia. But the Out-of-Africa story needs more evidence from this whole segment of Africa over the last 100,000 years. That which I have just given is not very compelling but something like it is to be looked for.

To sum up the African story, omitting North Africa, we see:

- Earliest moderns are found in East Africa, south of the Sahara.
- In southern Africa (with Border Cave), long occupation by one stem of moderns.
- In much of central Africa, a degree of very late homogenization due to the Bantu spread.
- Before the Bantu, signs only of late local divergences in Sub-Saharan Africa.
- Nothing to dispute Out-of-Africa migrations.

Chapter 17

The Mysterious East

This chapter is about the Mongoloid peoples of the whole eastern half of Asia. By themselves, they are a particular and important part of the problem of modern origins. Before clearing the decks for that one, let us take a moment to deal with some of their neighbors.

Little Dark People

The Negritos of the east are the short, woolly-haired hunting groups found in several of the Philippine Islands, from northern Luzon down to Palawan. There are legends and possible signs of such people in Taiwan. Southward, somewhat similar groups are found in the mountains of the Malay Peninsula. In fact, such a stratum suggests itself in eastern India and Ceylon. Finally, true unadulterated pygmoid blacks inhabit the Andaman Islands, stringing southward from Burma in the Indian Ocean. They do not at all resemble Australoids as we know them.

The Andamanese immediately recall Africa in skin, in hair, and in the Hottentot trait of steatopygia, or localized fat on the buttocks. However, they do not at all resemble African Pygmies of the Congo forest, and repeated studies by multivariate distance analysis ally them only occasionally with Africans. The Andamanese are mysterious and likely long to remain so; it is not easy to use them to argue for a direct Pleistocene connection of Negroid Africans with the east. Few scholars seem impressed with such an idea.

In general, excluding the Andamanese, many of these darker, shorter folk seem like an extension, or a remnant, of the Australo-Melanesian main branch of humanity. At least, that is where some Negrito crania place themselves when tested mathematically. Similar very small people live in the New Guinea mountains; here their size seems to be an adaptation for saving energy in constant steep climbing with burdens; otherwise they are entirely like other New

Guineans. In some islands of eastern Indonesia there are Melanes-
ian-looking natives and also, here and there, a few languages of the
New Guinea Papuan complex. Numbers of people from Indonesia
and the Philippines have bushy hair, and in general there are
suggestions of admixture from now-gone Negritos or Melanesians.
We accept that the Australoid peoples formerly occupied Indonesia
and part of Southeast Asia, and there is some direct evidence, like
the Niah Cave skull, that such was the case. However, the solution
is not essential to the main problems of Asia, and we can leave these
trace people to future study.

In modern India, signs of such a stratum exist, almost submerged under the vast population of Caucasoids (now dark-skinned
— see footnote 4, page 189) entering from the northwest. A late
constituent of these were the Aryans, speaking an Indo-European
language. (The name "Aryan" has the same origin as "Iran".)
Caucasoids also occupied much of Central Asia. In recent millennia
— all of the time since the pyramids of Egypt were built — there
has been pressure and invasion from the East, by Mongoloid nomads
familiar to history as Huns, Khans, and the like. These themselves
were also in conflict with the emerging Chinese. There is something
of a parallel with Rome, whose expansion in Europe spread and
established Western Civilization. Once "Eastern Civilization" was
formed in Bronze Age China, and China was united under the Han
Dynasty, Chinese rule and culture expanded, mainly southward.

Mongoloids North and South

Anthropologically, this is reflected today in the physique of two
kinds of Mongoloids. In the south and offshore are the varied but
essentially Mongoloid peoples, usually light of build and brownish
of skin. I would call them "mild Mongoloids."[1] What are thought
of as typical, or "classic" Mongoloids are an intensification of the
pattern, with the flattest faces and the most marked folds of skin
over the eyes. This pattern increases going from Chinese to Japanese
to Koreans to tribal peoples of eastern Siberia.

---

1. In one sense these could be called "proto-Mongoloids," which, however,
emphasizes a sense that the classics are more "typical." These southern populations
are not "proto" anything, being perfectly good Mongoloids.

These northerners are seen as an evolutionary response in relatively recent time to the intense cold of the latest Pleistocene. It is the southerners, from whom the "typical" northerners apparently arose, that pose the question, in this core problem of Asia.

In distinguishing broadly between the two great categories, studies of teeth by Christy Turner, of Arizona State University, are of primary importance. For about three decades he has been defining particular variations in teeth, like shovel-shaped upper incisors, a sixth cusp on lower molars (the dryopithecus pattern has five), the absence of third molars, and others. He has compiled the occurrences of about thirty such traits in samples from all over the world, computing statistical distances, like those using measurements, to group the samples. He has found that Asiatic peoples of the south and offshore share a degree of dental complexity, registered in such traits, that is absent in Australoids, Africans, or Europeans. He calls this pattern Sundadont (after the Sunda shelf and islands composing Indonesia). It is found in all those "mild Mongoloids" from Southeast Asia and Indonesia up through the Philippines to the aboriginal tribes of Taiwan and the Ainu of northern Japan, and out into the Pacific.

Northward on the continent, however, he recognizes a still greater level of dental complexity, which he terms Sinodont. His colleague in such work, Kazuro Hanihara of Tokyo University, has called the people in these two perceived strata "Archaic Mongoloids" and "Neo-Mongoloids," respectively, suggesting that the former are the basis of the latter, that is, that the two are stages of the same development. On the largest scale, what we have is a difference within a likeness. In any case, it seems evident that the roots of sinodonty are in the north, where it developed as an intensification of sundadonty, and that where found elsewhere it reflects recent outward movement of the people in whom it is expressed. We do not know why the pattern developed.

Gordon Bowles, a ranking authority on Asia, quantified the whole spectrum, using the generalized distance methods already described, applied by him to large numbers of measured samples of living people. Very grossly, he found a south-to-north zonation like the following:

- Branches of Turner's Sundadont people (from Burma around to Taiwan).
- Tibet and adjacent parts of China.
- East, west, and south China up to a region between the Yangtze and Yellow River basins.

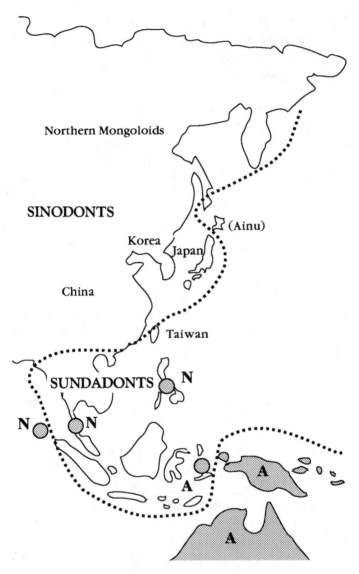

Fig. 61. The Distribution of Asian Peoples

Shaded are the remnants and populations of non-Mongoloid people, appearing as N (Negrito) or A (Australoids of Wallacea, Melanesia, and Australia). The latter peoples comprise the present aboriginals of Australia and Melanesia, as shown; the interest here is their presence as remnants. "Sundadonts" comprise Southeast Asiatics and offshore peoples from northern Japan (Ainu) down to Taiwan and Indonesia, as well as other Pacific Islanders. "Sinodonts" comprise the mainland populations of Korea (also Japan) and China, as well as peoples to the north and west.

- East Asia (North China, Japan, Korea), which he called "Intermediate Populations."
- Caspian Sea to Manchuria: Kalmucks, Kazakhs, Buriats, Mongols.
- Siberia and northern woodlands, reindeer herders, hunters, fishers: Tungus, Koryak, Chukchi, Eskimos, among others.

As we move down the list, heads and faces tend to get broader and faces higher, and Mongoloid features of eye and face become exaggerated. Most of the peoples of the last two groupings are Ultra-Mongoloids. The argument is compelling that, in this northern region, adaptation to intense cold during the later Pleistocene produced the flatter, broader face and fat-protected eye. Especially skeletally, the nose is narrow and flat between the eyes. This, then, is to be looked on as the core of the "classic" Mongoloid configuration (actually less extreme in Chinese or Japanese) of the great populations of the east.

The Varied Sundadonts

Back in the days when the races were popularly called the White, Black, Yellow, Brown, and Red, the people of Southeast Asia, Indonesia, and the Philippines were the "Browns." They vary physically, of course, but in this region it would take a most experienced eye to place a given individual closely to his point of ethnic origin. The fringe continues north. The aboriginal tribes of Taiwan are especially varied, with one of them, the Bunun, being particularly dark and short. Still farther north, before the arrival of the Japanese proper, the Japanese islands formerly had an indigenous Sundadont population, the Jomon people, whose only survivors are the Ainu of the north.

These last, popularly known as the Hairy Ainu because of their beards and body hair, are un-Mongoloid in looks. In fact, pundits of the past liked to think of them as "archaic Whites" or else as northern Australoids. But their skulls do nothing to support such hypotheses; instead, multivariate analyses show their affiliations to be with Asiatics of the Sundadont kind, above all with the Jomon people. The Ainu language has distant affiliations with some other north Asiatic languages, including Japanese. Naturally, we do not know the languages of extinct Jomon tribes but their relations were probably similar.

Why are these fringe people in fact on the fringes? Were they pushed? It is not likely. In the late Pleistocene, 10,000 years ago and earlier, they originally walked on dry land to Japan, Taiwan, and all of Sundaland (western Indonesia), where there are plenty of archaeological remains to show it. In many of these places we assume they found Negrito or Australoid groups already in residence. Good watercraft soon became part of the arriving cultures: language analysis shows that the outrigger canoe was present at 5,000 B.P., and watercraft are probably far older than that. But a major reason for expansion and population growth of the fringe people evidently had to do with food.

Europe, we saw, was invaded by grain-planting agriculturalists, but the origins of Asia's great native grain staples — rice in the south, millet in North China — are still being worked out by specialists. Probably even earlier, and well suited to the damp tropical climate of the Southeast Asian fringe, were foods propagated by cuttings rather than by seeds: yams, taro, plantains (bananas), and breadfruit. To these were added pigs and chickens, the latter descended from the Indonesian jungle fowl. All this indicates, in addition to the Near Eastern and Chinese centers of grain-growing, a third center of independent domestication on the South China shore and in the islands. The Sundadont populations would by this time have arrived in Indonesia, where a few hunting peoples of the older stratum still remain.

It is hard to tell just where and how these several domestications happened. For example, the world's earliest pottery goes back 12,000 years to the Jomon people of Kyushu, Japan, and continues on from there (the name in Japanese — Jōmon — describes one main kind of their pottery). Although apparently cultivating a few vegetables, these people were not growing grain, so it is not clear what the pots were for; cooking or storing water does not look like the whole answer. A couple of tooth experts think they know. Loring Brace at the University of Michigan finds Ainu teeth to be remarkably small (though skulls are actually large) and argues that long use of domesticated food allowed teeth to diminish by a kind of natural selection; he sees the same for the Chinese. Christy Turner is equally venturesome: recording a lot of cavities in Jomon teeth, he blames this on a starchy diet, namely taro. A little direct evidence might clinch the case; unfortunately the recent Ainu were not growing anything like taro.

It does seem odd to think that such tropical crops might have been at home in Japan. Elsewhere, however, there are other early signs of cultivation, as in Taiwan. In many southern parts of the Asian fringe, signs of very early gardening would now be submerged because of the post-Pleistocene rise in sea level. Nevertheless, the root-fruit-pig-chicken complex is so widespread that, regardless of later domination by rice, it must have existed as a whole economic world during much of the time since the Pleistocene ended. It still constitutes the food basis of various of the fringe people.

The history of the expansion of the fringing cultures is apparently reflected in language, in a parallel to Europe. Except for the Ainu and New Guineans, the offshore aboriginals of the whole Pacific speak languages of the Austronesian family. This family is much like Indo-European in variety, expanse, and probable age. There are no Austronesian speakers on the mainland — except for an insignificant patch on the Vietnamese coast — but at a higher level, the family has distant relatives in Asia, especially Thai, and it must have been mainland in origin. The most diverse subgroup consists of the languages of the aboriginals of Taiwan. Most students therefore believe that Taiwan is the oldest surviving seat of Austronesian and the point from which it spread down through the Philippines into Indonesia and out into the Pacific.

The Oceanians

Austronesian languages are spoken along the New Guinea coast and in Island Melanesia, where they have their own degree of special variety. Whether the languages and the taro and yam staples arrived together is not clear. But here is a most interesting matter. The great bulk of the numerous people of New Guinea speak entirely different languages. They are loosely called "Papuan"; in fact, New Guinea has by far the greatest linguistic variety of any part of the world. Papuan languages form a hierarchy of families, stocks, and phylums. By comparison, the Austronesian languages show only a low level of diversity.

Purely as an estimate, this diversity means that, from an original parent, Papuan languages have been diverging for at least 10,000 to 12,000 years. Furthermore, it is very likely that, in the long time after settlement at least 40,000 years ago, New Guineans independently arrived at food domestication, though it is not clear what crops were involved — perhaps a native banana and some edible

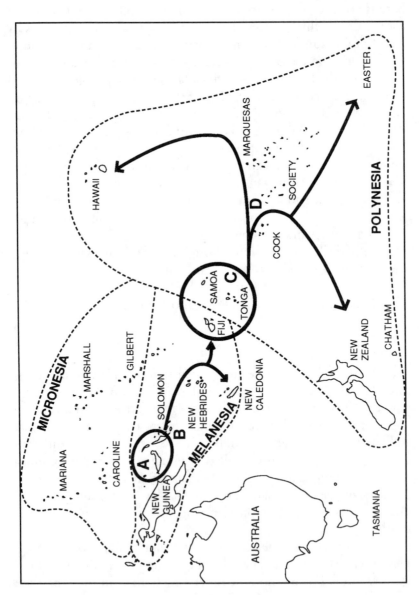

Fig. 62. Migrations in Oceania
A: to ca. 2,000 B.C.; B: 2,000–1,500 B.C.; C: ca. 1,500 B.C.–A.D. 0;
D: from A.D. 0.

greens. Digging in the highlands has revealed drainage ditches 9,000 years old, strong evidence of cultivation of something like taro or yams. Pigs also appeared fairly early — Jack Golson says at 8,000 B.P. — and pigs and taro would both have had to be introduced from Indonesia. Such early dates for these introductions are rather puzzling. In any case, interior New Guinea developed a heavy population well before recent times, and everything points to a long period of occupation and isolation.

Pockets of Papuan languages survive in Melanesia as far east as the central Solomons. If they went further, for example to distant New Caledonia where the natives look both Melanesian and Australian, they have been completely submerged by Austronesian tongues. This matter, of pre-Austronesian settlers in eastern Melanesia, is one of the pleasant conundrums the Pacific holds for future anthropologists. As to New Guinea, however, it seems clear that the advancing culture of the Austronesian-speakers came up against already established gardeners on that island, who may have adopted some things but were otherwise little affected.

Of particular interest is the deployment of the Polynesians. This has been well and lovingly studied. It is marginal to the story of our evolution, and we will give it only a hop, skip, and jump. And that is about what the Polynesians gave it, in getting to the eastern Pacific: New Zealand, Easter Island, and Hawaii. Polynesians all speak closely related Austronesian languages, as Captain Cook at once perceived. And they are closely akin to one another physically and cranially. But they do differ somewhat from people of the western Pacific: they are large in body, with notably big heads. This is most probably the result of a little natural selection acting on their recent ancestors. Philip Houghton suggests that the reason was the constant cooling breezes that blew from the open Pacific over the islands of Polynesia: in spite of the tropical location, loss of body heat could be a serious matter. Larger bodies have a relatively smaller surface-to-bulk ratio, so that heat preservation from available food is better. But this size difference makes difficult the search for Polynesian ancestors.

So it is not known whence the original Polynesian stem took off; it could have been anywhere from Taiwan to eastern Indonesia. In any case, by about 2,000 B.C., and probably in the Admiralty Islands north of New Guinea, the pre-Polynesians had improved on the general boatcraft and skilled navigating of the whole western Pacific. They began colonizing and trading voyages eastward through Melanesia, where their many small coastal settlements have been located archaeologically, detectable by their remains of Lapita-style pottery and of obsidian traded from New Britain at the western extremity. By 1,500 B.C. they had reached the end of the line, of such already substantial voyages of several hundred miles, at the triangle of Fiji, Tonga, and Samoa. After developing the definitive Polynesian culture just before the Christian era, while the Vikings were still skulking in the fjords of Scandinavia, they finally shot out in great ocean voyages, bringing themselves and their plants, pigs, and chickens to the center of Polynesia. Over the period between 400 A.D. and 1,200 A.D., they moved out again from the center, settling Easter Island, Hawaii, and New Zealand.[2]

The Center

Coming ashore again, we return to the "classic" sinodont Mongoloids of the Far East. They are like the Sundadont peoples but more accented: lighter skin, straighter hair, broader, flatter and higher faces, and a more marked eyefold. This is the basic character, with intergrading and overlapping in many areas. Nevertheless, repeated cranial studies using generalized distances by Michael Pietrusewsky of Hawaii, Loring Brace of Michigan, and myself tend to separate Sundadont and Sinodont samples and distinguish both of them very clearly from Australo-Melanesians.

As to history, today's Japanese arrived in their present islands from Korea, beginning just before the Christian era. We see this archaeologically in the Yayoi culture, bringing wet rice agriculture, a new kind of pottery, iron, and eventually horses. There is a considerable literature on the replacement of the Jomon indigenes, concerned with whether this was military or peaceful, and whether or not the Japanese themselves are largely descended from such aboriginals. Never mind: the evidence of skulls and teeth has spoken. The Ainu and the Jomon people, we saw, go with the archaic Sundadont peoples. Yayoi skulls, on the other hand, go with modern Japanese, and also with some peoples of inner Asia. And Japanese culture and language suggest an origin among the Tungus, who are

---

2. Why did they stop there? Doubtless they did not, but further voyages induced by past success would have one of two outcomes: loss in the great distances of open ocean, or inhospitable welcome by already civilized people of Central and South America.

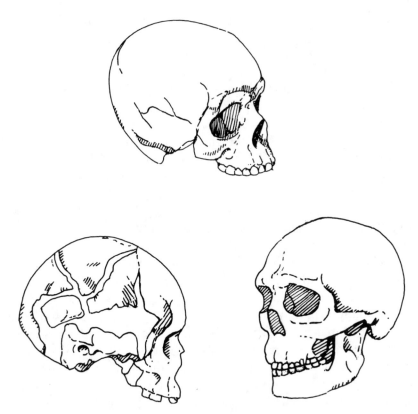

Fig. 63. Late Pleistocene Skulls From China
Top, the Liujiang skull from southern China, dated to 67,000 B.P. or earlier. The brows are more strongly developed than in typical Asiatics; otherwise, the specimen is entirely modern. *Right*, Upper Cave 101, dated to around 25,000 B.P. This rugged individual is hard to place as "Mongoloid." *Left*, Upper Cave 102, less well preserved, is flatter in the face and thus more Mongoloid in appearance.

widespread tribes of eastern Siberia. In any case, the Japanese are late-arriving classic Mongoloids.

So are the Koreans, who are facially even more "classic." Their language, although unique like Japanese or Ainu, is believed to place itself closer to the Altaic languages of much of inner Asia.[3] Not much is known about Korea archaeologically. There are signs of a parental

---

3. None of these is related to Chinese, which has southern affiliations.

kind of Yayoi pottery and suggestions are that arrival on the peninsula of the Korean people was relatively late.

The Chinese, by contrast, have a majestic history and prehistory, with written sources going well back to the first dynastic states before 2,000 B.C. All this history is northern. With the consolidation of China under the Han dynasty, beginning in 206 B.C., the "Han Chinese" with the dominant Chinese culture began the southward pressure that has continued to the present. Except for the "Minorities" (non-Chinese tribes in the south), China has incorporated or absorbed everything to the borders of Indo-China as well as, long ago, colonizing the islands of Taiwan and Hainan.[4] So there is a kind of spectrum from more to less "classic" Chinese from north to south. And, cranially speaking, people of the south tend to converge on Taiwan aboriginals (Atayals) or Filipinos.

There is not much evidence of history. Judging from limited numbers of Neolithic skulls (early farmers), the northern Chinese of today appear to have been in place 7,000 years ago. What we need is earlier examples of fully modern people. Two crania come to witness.

One is that from Liujiang, South China, found in 1958 in a cave filled with a few other bones which, if belonging to the same person, may suggest a female, although the skull in fact appears male. This extremely important skull has to carry a lot of weight. It is entirely modern except for a rather strong bony brow. A few writers have seen Australoid affinities in it but most, including Chinese authorities, view it as a proto-Mongoloid, what we are here calling a Sundadont. It has a few cranial features that are arguably Mongoloid. Its teeth are rather too worn for a full diagnosis by Christy Turner; however, it congenitally lacked wisdom teeth, something more common in Mongoloids than other main populations. In one multivariate analysis it actually comes to rest among Europeans, with the next nearest people being aboriginals of Taiwan. In another such analysis it comes down among Jomon and Ainu crania.

Its date, by uranium series dating, is set, not very confidently, at 67,000 B.P. The significance of that should be apparent: once more, we have a full-fledged modern skull, except for a touch of archaism in the brows, apparently foreshadowing its geographical

---

4. Recent tensions on Taiwan rest partly on the differences caused by an ancient immigration from South China and a recent one by the Nationalists of the north. The two groups spoke mutually unintelligible dialects of Chinese.

successors, and coeval with the late Neanderthals of Europe and the Levant, as well as with the early Australian settlers. If the date is correct, it lies in time between Omo 1 and the South African moderns on one hand and the Cro Magnons of Europe on the other. The bones of the skeleton are not large but are thick in the shafts, an archaic feature recalling Irhoud in North Africa or the Klasies River Mouth parts in South Africa.

The second specimen is the Old Man of Zhoukoudian. He has nothing to do with Peking Man and Locality 1 — except that his remains disappeared with the rest in 1941, leaving us now only with plaster casts. He comes from another deposit at Zhoukoudian, the so-called Upper Cave, where four skulls and other bones were found, dated to about 25,000 B.P. or earlier. The three best skulls have been repeatedly revisited: Weidenreich thought they suggested Caucasoid, Melanesian, and Eskimo affinities, respectively. The best one, the "Old Man" (Upper Cave 101) has also been assessed as approaching American Indians in character. Chinese anthropologists, as well as some Multiregional Evolutionists, have preferred, not very happily, to put him in a line leading from *Homo erectus* to modern Chinese.

This was the only skull of the three in a condition good enough for metrical analysis (based now on casts), and it has been subjected to intense scrutiny of this kind by David Bulbeck, by J. Kamminga and R. Wright, by Loring Brace, and by myself. In every case it affirms that it is no typical Mongoloid, falling generally in a space by itself between Europeans and American Indians, though statistically it is rather distant from anybody. (Brace finds UC101 to be closer to Polynesians than to Europeans.) At the same time, Turner pronounces the teeth of all three specimens (or rather, what can be seen of the teeth in the casts) to be sinodont, with North Asiatic implications. Unlike the case of the Cro Magnons, who were identifiably Caucasoid, UC101 cannot be grouped with the first farmers of his own region, whose skulls are recognizably Chinese. However, another of the Upper Cave skulls, UC102, is definitely more Mongoloid in appearance. Altogether, the Upper Cave situation brings us to another problem.

The Americas

The problem is this. The natives of the Americas came across the Bering region from northeast Asia. That is Ultra-Mongoloid territory today. The American Indians are not Ultra-Mongoloid.

Why not? Can this tell us anything about time and recent evolution in North Asia?

Indians have various Asian characteristics, along with differences (more on that in a minute). Importantly, Christy Turner finds all their teeth to be markedly sinodont, which is the stamp of North Asia (and, let us remember, apparently of UC101). The North American Eskimos, of course, are indeed strongly Mongoloid in form and are culturally and physically continuous with Eskimos, Chukchis, and similar peoples in Siberia. They arrived and spread across the Arctic a few thousand years ago.

The question of earlier migrations and movements is much debated, by first class students with excellent materials. Christy Turner, using teeth, Stephen Zegura, using single-gene traits, and Joseph Greenberg, the eminent linguist, have together proposed a scheme of three migrations, from different probable hearths in Northeast Asia. Eskimos and Aleuts were the last of these. Next to last were NaDene speakers, mostly the Athabascans of the Northwest, also including their kin in the southwestern United States, the migrant Navajo and Apache. NaDene languages may have connections with the Sino-Tibetan family. This migration is placed a few thousand years before the Eskimos.

The earliest hypothesized migration is simply All Other — characterized as Amerind linguistically, Paleoindian archaeologically, and varied and rather nondescript physically. Greenberg is willing to put the many divergent languages of the Americas into a single great stock whose first divergence would reach back beyond 10,000 years, to a point where most methods of tracing relationships give out. Authority that he is, some scholars find Greenberg's breathtaking consolidation pleasing and thought-provoking; others are discomfited.

The scheme of three migrations should be taken as a logical approximation, not as settled truth. Objections, major and minor, can be made on closer approach. Two able Canadian women, Emöke Szathmary (who works with genetic traits) and Nancy Ossenberg (who works with many specific cranial traits, like Turner with teeth) are quite unwilling, after intensive analysis, to sort the northernmost peoples of North America so neatly. But it is the rest of the American populations that we should attend to.

In particular, work with skulls confirms that these populations are unsystematically varied and do not tell any clear story. Like Asians they tend to be broad and flat in the face; unlike Asians their

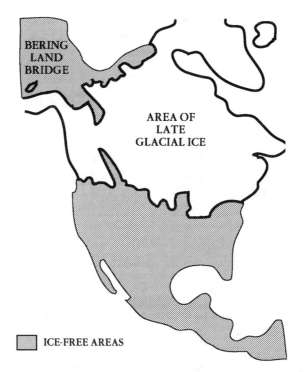

Fig. 64. The American Glaciation and the Bering Land Bridge
This shows the connection between the western and eastern ice sheets of
North America, before a corridor opened between them about 12,000 BC.
In the northwest, most of Alaska and the exposed floor of the Bering Sea
formed a wide land bridge from northern Asia, a bridge which was itself
habitable to entrants from Asia.

upper noses are quite prominent, their foreheads are flattish, and
their skulls are decidedly broad across the base. Also their differences
in appearance from nation to nation are probably familiar to many
of us. In the broadest studies of cranial "distance," Indian groups in
fact ally themselves almost as often with Europeans[5] as with Asians;
and attempts to find likeness with specific Asian peoples are not

---

5. Readers may have noticed that in low-budget Hollywood movies American
Indians are readily represented by Caucasoids with a little face paint. The
magnificent Lakota actors in a high-grade film like "Dances with Wolves" are
something else.

successful. So this general apartness gives us the chance to inquire, in terms of time and places of origin, something about developments in Asia itself.

Further, the resourceful Marta Lahr has analyzed some of the rare skulls from southernmost South America (Patagonia and Tierra del Fuego). She finds the kind of large size and robustness seen in late prehistoric North Africa (Afalou), and to a degree in Australia. This robustness might be appropriate as an adaptation to a cold climate, but it would also be appropriate to an advance guard of Indian arrivals marked by the kind of robustness seen in late Asiatics like the Upper Cave folk at Zhoukoudian.

Accordingly, the crucial thing becomes the time of arrival of these First Americans. It is much under dispute. Not in dispute is the outwelling, after 12,000 B.P., of the Clovis Big Game Hunters. By their characteristic, beautifully made stone projectile points, they are traced all over the then ice-free parts of North America right after their first appearance. And many surviving large mammals of the Pleistocene — mammoths, horses, giant ground sloths, and others — promptly disappeared at this time. In fact, one authority named this phenomenon the "Pleistocene Overkill," picturing the Pale-oindians as advancing rapidly south and east in a mere thousand years, proficiently killing everything they met. Whether things are quite so simple is immaterial to discussion here;[6] the important matter is the rapid occupation right down through South America.

The further questions are: 1) how did these immigrants come, and 2) were they indeed the original comers? For the first question, there can be no doubt that they crossed Beringia, which was then, as periodically in the past, not a strait between America and Asia but rather a broad plain, exposed by late Pleistocene low sea levels and abounding in game animals. Like much of Alaska, it was not ice-covered and it could have been entered from Asia during most of the late Pleistocene, when it would not have looked like a narrow connection at all.

So, for thousands of years before 12,000 B.P., crossings could have been made into Alaska, where such early settlements are claimed. But getting into the rest of the continent was something

---

6. As already noted, such an extermination of large animals with the arrival of people would have some parallels elsewhere: New Zealand (moas), Madagascar (giant lemurs), and Mauritius (dodos).

else. The huge North American ice sheet was joined to that of the western mountains. The lock was broken — a corridor between them opened, or else the travelers took to boats — as the Pleistocene ended, and the Paleoindians swarmed south. Just now, however, antecedents of their Clovis point culture, whether in America or Asia, are not known.

Question two: were they the first? Any earlier people would have had to be much earlier, or else to have come by routes unknown. There is a long history of alleged evidence of early Indians. One by one the claims have been dispatched under closer examination because of innocent mistakes or faulty evidence. But the claims keep coming. Today, such cases are put forward not by amateurs but by first-grade professionals, and the disputes are warm. Recently found sites have provided good dates of almost 20,000 B.P. in Pennsylvania and over 30,000 B.P. in Brazil. The peculiar thing (at least to me, a doubter) is the isolation of such spots. After 11,000 B.P. the Paleoindians were everywhere, as were ancestral Europeans and Australians soon after their own local arrivals. No such expansion or cultural continuation is seen from the few alleged early American sites. Also, they betray no Clovis connections. It is all very puzzling.

Answers, positive or negative, will be important, for reasons having to do with Asia and the Mongoloids generally. I said earlier that the Mongoloid peoples are a world to themselves. They vary, and the American Indians seem to make the furthest deviation. Nevertheless, apart from cranial and facial signs of general kinship, there are some special marks of Asian-American commonality. Teeth are one: the pronounced sinodonty of the Americans is significant. Another case is a particular blood antigen, the Diego factor, which occurs at moderate frequencies throughout the Americas and at lower levels in Asia. It is not met with in the Pacific (except the Marianas where there have been recent immigrants from the Philippines) or elsewhere in the world. And a special form of ear wax — crumbly instead of waxy — is the prevailing form in northern Asia, with lesser proportions in other Mongoloids, including American Indians and Pacific peoples, especially Austronesian-speakers. It is extremely rare in Europeans and is not known to occur in Africa at all.

So knowing the time of separation of American Indians from other Asiatics, in Asia itself, would have an importance. How would this relate to the Upper Cave "Old Man," also pre-Chinese but non-Asiatic in character? Would this suggest that the whole "classic"

Mongoloid physical development is quite late? How does it bear on recent *Homo sapiens* generally?

The Middle of the Muddle

The still more fundamental question would seem to be: where do we get our Sundadonts, varied but similar, non-robust, of a basic Mongoloid character? From Australoids of Sundaland? Hard to swallow. From regional evolution from *Homo erectus?* Also unpersuasive; and fossil evidence from Mapa, Dali, and Jinniushan is unconvincing, with only Liujiang as an acceptable possibility. Migrants from further west? No signs of Levantine connections, like Skhūl-Qafza or early Upper Paleolithic folk.

Christy Turner has a broad scenario, suggesting that in fact the Sundadonts are the source of most of the "racial" groups of recent times — a sort of "Out-of-Southeast-Asia" plot for perhaps the last 50,000 years, following the Out-of-Africa movements of earlier times. From his extensive bank of data on teeth he calculates statistical distances among many populations, and finds that Southeast Asian groups seem central, with the lowest distances to others. (And I have found that such people have the most "average" skull shapes among moderns.) Those seemingly too distant — statistically as well as geographically — to be offshoots are American Indians and some North Asiatics (very strange!) as well as Europeans and Africans. His student J. D. Irish has used the same studies of teeth on Africans, finding their pattern different from sundadonty, and marked by archaic traits, which sundadonty is not — another signal of African antiquity south of the Sahara.

We should remember that, while this broad-scale evidence of teeth is very good evidence, it is essentially from living peoples, like molecular evidence (with possible rare exceptions), and so we are looking at distributions, not direct ties to the remoter past.

Turner finds Australians within the acceptable distance for sundadont parentage. This brings us right up against basic questions. Australian crania are robust; those of Sundadonts are not. We know now (since Turner wrote the above) how very early the Australoids arrived in that continent, and how they must have been in Sundaland beforehand. Given this, could Sundadonts as we know them have given rise to the Australians? Contrariwise, could Australians have evolved into Sundadonts in Sundaland? Just now, we lack the key to Southeast Asia.

Chapter 18

The Big Picture

At the beginning of the twentieth century, scientists were perfectly clear that we had evolved. But there were almost no fossils — Neanderthals, Cro Magnons, and the then and only Java Man — and dating was a matter of despairing guesses. At the dawn of the twenty-first century, serious gaps remain, but they are well-located gaps. We have really impressive numbers of human fossils, and the structure of time is clear. It will be interesting to see the state of affairs a few more generations down the road.

The World Is Round, After All

After tracking events continent by continent, we now try a global perspective. Here I use some recent data-based analyses on living peoples; many earlier attempts have been simple flights of the imagination. The data are of two kinds, already described: cranial measurements on the one hand, and "single-gene" traits of blood, enzymes, and proteins on the other. With statistical methods these data will allow populations to be gathered into grand clusters that may look better than they are. I present graphs of two such clusterings for comparison, one cranial, using my own material, and one genetic, the production of L. L. Cavalli-Sforza of Stanford and his associates.[1]

---

1. The cranial clustering is based on 57 measurements made on 28 sets of male skulls each highly specific as to location and community of origin. The reduction aims at translating the measurements into general skull shape, holding size constant. Little is known about influences acting on shape, and so it cannot be said how much the divergences of peoples owe to history and how much to environmental effects; indications are that history is well reflected.

   The genetic graph uses 120 separate genes tested for in 38 population groups. Limits on the conclusions that can be drawn from gene testing have been

The cranial graph shows only populations for which samples of skulls were available (no peoples of India, no Koreans, etc.) It makes expected main separations by geographic area. The first split sets off Africans, but this branch also contains Australoids. The opposing main branch contains all others. One sub-branch allies Europeans and American Indians. The other such sub-branch is all "Mongoloid." Its first division contains proper Asiatics: Japanese, Chinese, Taiwan aboriginals, and Filipinos; its second division is made up of Mongoloid irregulars: Polynesians, Ainus, Guam Micronesians, and Eskimos. This is all illuminating and makes sense, except for the puzzling Andamanese.[2]

The graph of genetic factors separates Africans still more sharply from all others. Elsewhere, the interesting feature is the proximity of Caucasoids to North Asians and American Indians and the distinction of these from South Asians and Pacific Islanders; here Australoids are loosely linked with the last.

There is more agreement than disagreement between the two approaches. A major matter is time. My own conclusion, from the cranial evidence, is one of recent separation of all peoples. Similarly, Cavalli-Sforza et al. frankly believe that the recent replacement hypothesis (ROA) is upheld by their analysis. A later study by the same group, using nuclear DNA rather than blood genetics, persuades them still more strongly.

---

described earlier: some gene distributions are related to prevalence of malaria, for example, and are thus not good measures of gradual divergence; others are simply not very informative, while still others are highly so.

The multivariate statistical procedures are designed to distill the main outlines of likeness — or more accurately, of least unlikeness — among the groups included. Both graphings have here been compressed from the originals. This is done without remorse because, in such clusterings, exact results do not emerge; instead, the clusters chosen are simply the most likely of many possible solutions, depending on just what material goes in and how it is treated. However, these cranial clusterings are much more stable and mutually similar than the "trees" produced in mtDNA analyses.

2. In various runs of the cranial data, these mysterious pygmy blacks are sometimes allied, as here, with Africans, and sometimes not. They are not included in the genetic analysis, so that the same comparison is not possible. As we have seen, Greenberg placed their languages in an Indo-Pacific phylum, which would ally it with Tasmanian and the Papuan languages of New Guinea. This, in turn, would imply that the physical concomitant is "Australoid." The languages fall into two unrelated groups, which implies great antiquity for this isolated people.

CRANIAL CLUSTER
28 GROUPS

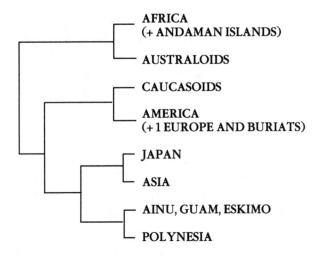

GENETIC CLUSTER
42 GROUPS

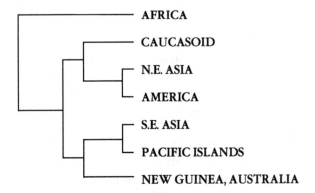

Fig. 65. Clusterings of Major Populations by Skulls and by Blood
These are the major branchings in "trees" having many more population groups in the originals. There is a good deal of basic agreement, especially in the first separation of Africa as a main branch, and in the joining of Europeans (Caucasoids, which includes, e.g., Egyptians) to Americans and north Asians (Buriats in the cranial tree). In another main branch, Sundadonts go with Pacific Islanders. In disagreement, Australoids (including Melanesians) are closer to Africans in the cranial clustering, closer to other Pacific and Southeast Asian peoples in the genetic tree.

Others are not fully convinced. It is to be noted that for neither method is there a calibration for time. That is one reason why Cavalli-Sforza and his group have introduced language in evidence. Using the latest stratospheric joinings of all existing languages, they find a very good correspondence with the genetic branchings. African languages (except Ethiopian and North African) are the most isolated. One kind of supergroup ("Nostratic," a contribution of Russian linguists) joins North Asian and Indo-European and, by a possible extension, American.

Does this bring us to any conclusions? Language change and divergence makes a non-rigid time scale. But the suggested joinings imply a first divergence that would be pre-American and pre-Australian, which would take us out beyond 75,000 years and more. At the most extreme, this would imply a single original family of modern languages, but few imagine that language relationships could be traced for anything like such a period. Cavalli-Sforza et al. are suggesting that complex language — of the kind used by all living peoples — was a factor in making possible the rapid worldwide expansion of moderns. They are not alone: recall that archaeologists have perceived striking new things as accompanying the "sapiens explosion": varied and refined tools, larger social groups, art and personal decoration, rituals of burial, etc. As the agency behind this they, the archaeologists, favor the appearance of a capacity for more highly developed syntax in language, bringing the construction and exchange of ideas to a new level.

Food for thought. Until recently it was believed that the Neanderthals were anatomically unable to make all the vowel sounds we use and would have sounded like our children, because conformation of the base of their skulls suggested a non-modern position of the larynx. But the fully modern hyoid bone (Adam's apple) found with the Kebara Neanderthal in Israel argues against that conclusion. And such an anatomical limitation has not been suggested for equally early people like Skhūl-Qafza and certain Africans. The Neanderthals were able, at the least, to copy Upper Paleolithic ideas. But copying is not inventing.

The Great Debate

In the meanwhile, the contest of the Multiregionals and the Replacementeers goes on. It has been an interesting contest for over twenty years. To resume, the two opposing hypotheses are clear in their central ideas. The Recent-Out-of-Africa explanation sees late exits of modern *Homo sapiens* from that continent. Local forms ("races") took shape mostly after populations had spread through the Old World, replacing survivors of any more archaic groups, if in fact any such survived. The Multiregional Evolution view sees a vastly earlier spread from Africa of *Homo erectus*, from which different regional samples rose vertically, though interconnectedly, over a million or so years to become their modern representatives.

Both have their qualifications. The ROA dispensation allows for some differentiation of parent lines within Africa, somewhat different times of exit for different lines and different histories of population contraction or explosion or population size. Theoretical studies of such factors are being worked out using mathematical models. The MRE view does not simply see long isolated differentiation in each area, independent of other areas. Instead it proposes continual contact and exchange of genes across the Old World, homogenizing the species, balancing the tendency of regional populations to diverge and passing useful adaptations back and forth. As we have seen, in one guise the proposition is known as the Single Species Hypothesis, holding that the continuum, horizontal and vertical, does not even allow for a species change from *Homo erectus* to *Homo sapiens*. Thus there is no place for a *Homo neanderthalensis* or for replacement of Neanderthals in Europe.

The Multiregional school faces various problems. The argument for separate local continuities, first made by Franz Weidenreich, has been disestablished by Marta Lahr who, in careful tabulations, shows that traits claimed for a local lineage are not exclusive to such regions at all.

Other assumptions are weak. The million-year interregional exchange of genes is not persuasive. Dog breeders, working with a single species, *Canis familiaris,* have expanded the variety already seen in Great Danes, barkless Basenjis of the Congo, Afghan hounds of the Near East, Pekinese of China, Chihuahuas of Mexico, and so on. Reversing this, the MRE view assumes the opposite: a delicate and balanced mongrelization over many thousands of years, to counteract evolutionary tendencies to diverge.

And this flies in the face of important events. In Australia it has been argued that the aborigines, one segment of moderns, descend from Solo Man while converging on other moderns. Contrariwise, it is clear that, at the other end of the Old World, the Neanderthals for several hundred thousand years were not following the same

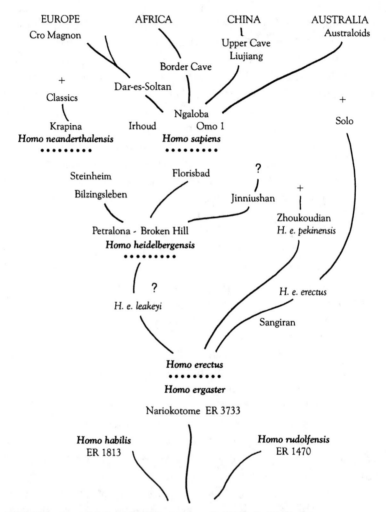

Fig. 66. Road Map for Homo or Splitter's Delight
This suggests the progression of species of *Homo* as set out in the text, using currently favored names. In particular it suggests two species arising separately from *Homo heidelbergensis*, in Europe (*H. neanderthalensis*) and Africa (*H. sapiens*) respectively. The usual designations are given for separate populations or subspecies of *Homo erectus* for Africa and Asia. The vertical spacing of names is approximate and is not intended to assert precise datings.

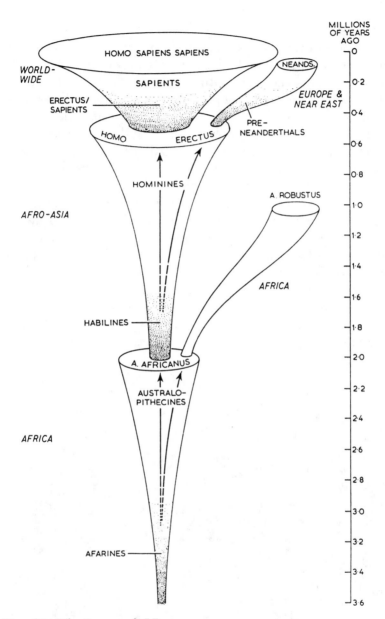

Fig. 67. The Lumper's View

This contains the same general information as the Splitter's diagram in Fig. 65, but the perspective is on the basic pattern of hominid succession, without much attention to variety within any stage. The figure was drawn up a few years ago by a leading authority, Michael Day, and it is not meant to take account of some recent problems, but rather to present a kind of overview.

script. Instead, they were steadily evolving away into a new species. Furthermore, over the last few millennia of their existence in Europe, there are no credible signs of gene exchange with Cro Magnons. True, some of the faithful argue that Neanderthals had a sudden conversion and developed rapidly into modern Europeans. But most workers can barely stifle their yawns at that one.

In the face of new information showing that the *Homo erectus* Ngandong population survived well into the time of the Australians, MRE writers have said that the continuity from the first to the second is nevertheless incontrovertible. However, as above, Marta Lahr has controverted it with her careful demonstrations that the perceived common features are far from exclusive to the two groups.

Here is a main point. The two approaches are not symmetrical. Multiregional Evolution is one general model with many exposed surfaces. It cannot be proved, but only disproved. There is little or no room for modification or disagreement. That is essentially the reason why the Neanderthals are such a problem.

Replacement Out of Africa, on the other hand, is additive and correctable. Adherents do not hesitate to disagree and have a variety

"WHERE ON EARTH HAVE YOU BEEN? WE ALL EVOLVED WITHOUT YOU."

Fig. 68. Still Another View

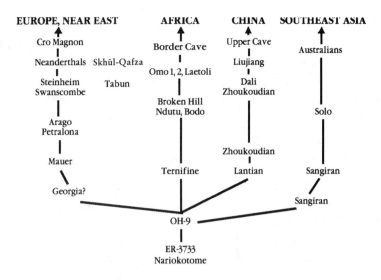

CANDELABRA (CONTINUITY)

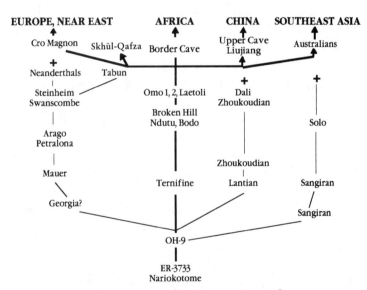

NOAH'S ARK (REPLACEMENT)

Fig. 69. Contrast Between Regional Continuity and Replacement Hypotheses

These are versions of a mini-pegboard: the fossils are fixed as to time and place, but the proposed connections indicate local descent on the one hand versus a relatively rapid dispersion, from Africa, about 100,000 B.P. or earlier.

of somewhat different plans. If the conjectured jump dispersal, sending African emigrants across South Asia to Australia should be disproved (how?) this would not demolish other parts of the story. Mitochondrial and other genetic evidence can be cited both for and against Replacement but not, as far as I know, in positive support for MRE.

March Past

What follows, admittedly a Replacement view, is only one general scenario, from which other ROA interpreters would certainly differ in detail.

As we look back, we envision several stages or events:

1. After about 35,000 B.P., with the Neanderthal and Solo populations gone, there survived no non-moderns or archaics.

2. From this general period back to 100,000 B.P. or earlier, three species of *Homo* were present: *sapiens, neanderthalensis*, and *erectus*. This carries some *Homo sapiens* back to a state with archaic trace survivals, during which dispersals of these and of early moderns took place from Africa.

3. About 130,000 B.P. or earlier, *Homo sapiens*, though pre-modern, is recognizable in Africa. Provisionally, we see descent from *Homo heidelbergensis* of Africa — let us say from Broken Hill to Florisbad — over the period following 400,000 B.P. This recognizes a real species instead of an "archaic" stage. Events in Asia are problematic.

4. Provisionally again, *Homo heidelbergensis* was also present in Europe. Thus there was a fork in the road, the second branch giving rise in Europe to Neanderthals, the first giving rise in East Africa to moderns. Elsewhere only *Homo erectus* was present.

Following is a closer reconstruction of late events, largely as proposed by Marta Lahr, partly with Robert Foley. Much of it involves "jump dispersals" or pulses of population movement, as well as phases of contraction and isolation, in which climatic events such as a wet or dry Sahara were important. The first phase:

1. Emergence of pre-modern *sapiens* (not present-day forms) in East Africa.

2. Migration from East Africa to South Africa (Border Cave, Klasies River Mouth, etc.) and to North Africa (Jebel Irhoud et al.), and probably to Central Africa. This would give a plausible frame

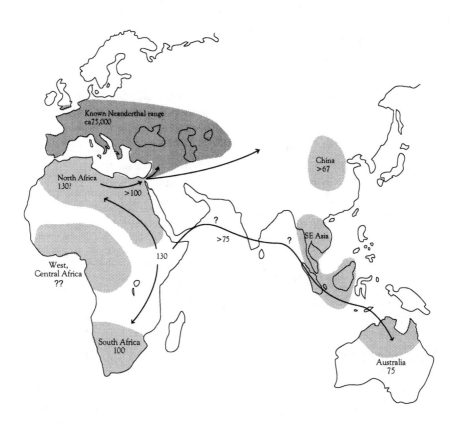

Fig. 70. Speculative Emplacement of Moderns at About 100,000 B.P.
This envisages earliest moderns in East and Northwest Africa at about
130,000 B.P. In what would be a frankly Out-of-Africa scenario, this
would be followed by: 1) a first impact on the Near East at more than
92,000 years ago; 2) an isolated population in Southern Africa indicated
by 100,000 B.P.; and 3) the conjectural, but logical, existence of a distinct
population in West and Central Africa.

The foundings of Chinese and Australoid modern populations are
contested: if from the west, the Australoids were probably early, by a
southern route. The Chinese were probably later. North Asia is not known
to have been occupied at all, by Neanderthals or moderns, before about
30,000 B.P.

for early San (Bushman), Caucasoid, and Negroid peoples as we know them.

3. Jump dispersals out of Africa: out of North Africa into the Levant (and elsewhere?) possibly more than once; and out of East Africa to South and Southeast Asia.

4. Later special events and localizations, viz.

> A) unknown expansions in equatorial Africa.
>
> B) Caucasoid replacement of Neanderthals in western Europe, and probable movement of Caucasoids eastward in Asia, as far as Lake Baikal.
>
> C) occupations of Australia and Melanesia from Sundaland.
>
> D) establishment of Sundadonts in East Asia, probably north and south, with subsequent development of North Asian Mongoloids and Siberian sources of American Indians.

This last list is virtually a confession of how little we actually know of the last 50,000 or so years, so close in time to ourselves.

What You Get Is What You Get

Geologists or astronomers generally find their revealing new discoveries in the directions in which they are already headed. Anthropologists are more like children not knowing what Christmas morning will bring but happy to get anything, even another sweater just like the one they have on.

Santa Claus has been good to them in recent years, particularly in the matter of dates. But what they might want is not necessarily what they get. We would certainly like more bread-and-butter information about the east of Europe and the west of Asia, for the time just before the appearance of Cro Magnons and the Upper Paleolithic. Much more about Africa for the same period; and of course the source and development of Sundadonts in the east. What we have, instead, is early dates for Australians and late dates for Ngandong. This is immensely informative but unexpected — a big surprise. Given today's knowledge this is about the biggest kind of surprise we will be facing — nothing like the finding, early in the century, that Africa, not Asia, was the home of hominids.

A wish list could be a long one. For example, a Miocene ape for the base of the hominid tree. More on those early australopithecines like *Ardipithecus*, complete with feet. Further fossil footprints, like those at Laetoli, would be too much to hope for, but they would show actual use, not inferences from anatomy. More on the *Homo erectus/ergaster* stage in Africa, early and late, with light on the problem of whether there are actually two species. Something more revealing about *Homo heidelbergensis:* is he real? And then much more, next year if not this, about emerging moderns everywhere, especially Asia. Without waiting for good luck, we will be going after information on the living ourselves, with things like DNA, since we do not need the fossils here.

Much of what I have written may strike more sober colleagues as speculative simplification. Of course it is — in a book like this, simplification is the name of the game. We now have masses of facts, often unconnected. Do the data speak for themselves? Says Milford Wolpoff, paladin of the Multiregional Evolution forces: "I have been in rooms with data, and listened very carefully. They never said a word."

So organization is the first step, though it may lead to such conflicting organizations as MRE versus Replacement, two interpretations drawn up from the same facts. Reading over the formulations and hypotheses put forth in past decades would show you how the ground can shift under existing interpretations, with the appearance of fresh information and with clearer interpretations of what is in hand. But only by continued efforts at organization can new ideas, dates, and fossils be grasped for what they mean.

Chapter 19

The View From Here

A hundred years ago, it was easy for readers of evolution to view us as the pinnacle of life, with other animals tailing off into more and more primitive forms. Evolution had marched straight ahead toward the human goal. It got there, and here we were, the finished product. This was a cheerful view, comfortable in the religious and philosophical furniture of cultivated Europe and America. Today things do not look so simple and pleasant. Of course we dominate the earth, and of course we are superanimals. But it seems paradoxical that evolution should have produced a species so hostile to the rest of nature.

And of course we are successful, at least among vertebrates — count our numbers. However, there are other pinnacles of life. If you want to view success, go to the insects. There are millions of insect species — ten million is a low estimate. So many are they that, if the naturalists can figure correctly, we are killing off more rain-forest insect species, before they can even be counted, than have already been recognized by science.

In other kingdoms, viruses and disease organisms can evolve so fast that they become immune to our "cures." We readily bring about the extinction of higher animal species, especially mammals and birds. But the extinction of smallpox was an extremely rare and difficult accomplishment, and the newly evolved AIDS viruses have probably settled in for a long future. If it is any consolation, when we ourselves eventually become extinct, we will take a lot of such forms of life along with us.

Did We Have to Be Human?

I hope I have shown that what formerly seemed like a steady, unblinking advance toward a state of perfected humanity now looks more like a series of lurches and lucky turns along the way. We have a good structural plan, that of primates, with excellent eyesight and grasping extremities. The ape branch became large-bodied and intelligent compared to monkeys but seems somehow to have fallen back catastrophically from its early success, certainly in numbers of species. Then hominids, becoming fully upright by some happy shift, started new experiments. But at least one result, the robust australopithecines, failed in due course, becoming extinct. We are not clear as to possible later extinctions. It does appear that in later times three species survived — *H. erectus, H. neanderthalensis,* and *H. sapiens,* until the first two also succumbed, leaving the field to us alone. Instead of a confident cakewalk to a natural goal, our arrival at humanhood looks more and more like a squeaker.

Ask a couple of questions that say something about our self-image. Are there other intelligent beings in space? Quite possibly. Do they look like us, upright, with eyes, nose, and mouth? Perhaps a little: a head equipped with such things is a sound anatomical design, as insects and vertebrates know. But, on the whole, if we glance back at the past we can see what a twisted path we have taken. Bats and seals, for example, evolved more straightforwardly: their limbs became modified for flying and swimming, without side turns. We, however, began as primates able to grasp things with all four limbs and then, as hominids, we managed to twist our spines upright and walk on the hind pair of hands, nicely rearranged for standing, striding, and sprinting. Are we to assume that space visitors might come from a world having recognizable vertebrates, mammals, and primates in their background? Hardly. Astronomers are agape at the surprises presented by the other planets and their moons. Kinds of life in farther space are equally unimaginable, and the chances must really be zero that arriving visitors will be anything like us.

So this is the first lesson. Evolution has no readable guidelines; it just happens. We do not know much about its limits. It has intricate possibilities. It tends, we have seen, toward more complex, more highly organized life forms, but it is not driven to these. We have seen how different classes of vertebrates were successively more advanced or complex. Still, once the mammals had arrived they were, so to speak, on their own, guided not by blueprints but by natural selection, life's master-builder.

And we do not see future openings for a still more advanced class of vertebrates. Primitive reptiles were visible among the early amphibians, and small mammals were present among the ruling reptiles, before these improving kinds of animal actually usurped the thrones of their parent classes. If there is something among the mammals corresponding to these earlier interlopers, it is the primates — and ourselves. In the previous usurpations, only fringes of the older classes survived. As a consequence of our own presence, it is not the nobility of other mammal life that prospers; it is the rats and the moles.

From Here On

A more important question: are we, the human species, still evolving? Of course — how could we help it? A lot of talk about this has been based on projections: brains getting bigger, heads getting balder, teeth getting fewer and smaller. Some of these trends can be seen here and there in the recent past, let us say the last 30,000 years. In Europe, we saw, Upper Paleolithic teeth were indeed larger and skulls more rugged than in their living descendants. Such change is likely to occur more rapidly in the smaller, more isolated populations like those of a few thousand years ago; it would be harder to detect similar change going on today, over all the vast numbers of living humanity.

A more sober and realistic view, one of possibly rapid evolution, would start from Darwin's central, essential point: who has the babies? In Darwin's terms a species evolves largely by natural selection. Individuals who have slightly more favorable adaptations will survive better. Surviving better, they will contribute more offspring to the next generation. Thus, insofar as their useful degrees of adaptation are "hereditary" (genetically determined), they are more apt to be inherited by the next generation. And so the population evolves in the direction of these individuals.

That is Darwinism, the way simple evolution goes in nature. But notice that the first part, about natural selection and adaptation, is not absolutely necessary to the end result. The species will evolve in the direction of the most offspring, period. There used to be a school of "social Darwinism" (not Darwin's idea) of a might-makes-right persuasion, that saw human success as a matter of rising to the top of one's generation through intelligence, ability, forcefulness, and, if you like, ruthlessness. These people missed the point. What counts is having children.

The human species — and remember that we are all one species — is far removed from a state of close interaction with nature. And recent centuries have made matchsticks of natural selection in our own case. Wind and water power, then steam and oil, have rendered strong backs almost irrelevant. Sailing ships broke the barriers of isolation between populations and between continents — goodbye to small-scale evolution in small tribes of people. Some populations have expanded in place, as in China and India. Others have been transported wholesale, like Europeans and Africans to the Americas, where they expanded afresh. Some aboriginal peoples were virtually eradicated, like those of North America and Australia. Although they still have their numbers, those people, who once held entire continents, will not figure in the further evolution of the species as a whole. So earlier balances have been upset.

You may write your own prophecies as to the future outward appearance of *Homo*. My own is that it will not be a smooth blend of the world's present people. Blending is likely to be overtaken by other kinds of population events. But the specifics of such events are so volatile, and so socially induced, as to make prophecy a foolish pastime. Volumes are being written on the subject, and this is not one of them. We can see the kind of thing that has been happening. We know that birth rates can shift rapidly. But the most dramatic shift has been that of death rates, especially infant death rates, at the hands of modern medicine; and since birth rates have not yet fallen equally, some parts of the world have become overstuffed with people. This is not exactly Darwinian but it is, shall we say, a fact of life.

A Question

Creation is the ultimate challenge to our imaginations. Astronomers, for example, can now describe and explain much of what they see in curtains of galaxies and the effects of black holes. Biologists, at the other extreme, perceive the smallest of the directors of life at work, and are progressing steadily in this pursuit. But there are more questions. The rules of physics and chemistry are the same in it all. Gravity has to hold throughout. We knew gravity would be less on the moon before we saw astronauts jouncing about on its surface. We can see the makeup of stars by the same spectroscopic analysis, first used here on earth, as telescopes get more wonderful: evidently, the same elements behave the same way, here and trillions of miles away.

The final questions are: How did these rules get set? Who established "Newton's Laws?" How did the ultimate particles of matter come into being, and how was it decided what forces and combinations of these would make up the elements, from hydrogen up, each with its given character? Such incomprehensible things must seem to us like real and conscious acts of creation, questions for which we have no answer.

Another kind of Creation, as in "Creation Science," wants to foreclose all such speculation, being content with Scripture. While it does not bear real scrutiny, it satisfies great numbers of adherents. The reason for the force of its hold is, obviously, a human question, and one that we can consider.

This leads into questions of human consciousness. Whatever that may be, it obviously has emerged in the course of our recent evolution. We have talked about the "sapiens explosion." We have much to learn about this, but it has to have been something real, separating us from earlier ancestors at some point.

Take language. We have forced laboratory apes to copy some of our ways of communicating. But they have no language of their own. We have languages, with common patterns across the species. Linguists are now convinced that human infants are indeed genetically endowed, born already wired to understand and use these patterns.

Behaviorally, apes and a great many other species of animals are social animals, and in fact are dependent on their societies. The evolutionary and genetic aspects of all this are developed in the emerging science of sociobiology, though this has its vehement detractors. However, the patterning of human social behavior, under the control of human culture, is greatly varied and adaptable. Again, we must acknowledge that our capacities for this came about through evolution. They can easily be considered a part of the sapiens explosion, though we can say nothing about behavior in earlier hominids, and just now had better not try. Although tightly biological social behavior might prevent the pathologies of modern societies, what we have evolved is probably highly appropriate for societies living in a stage of smaller and less pressured groups. The natives of Australia were living such a life, and were possessed of highly developed social forms. These related individuals to one another in totemic groups. This not only gave them their personal connections with nature and religion, but structured all their behavior: whom they should marry, and whom they could not marry, with all their social obligations clear. These could extend to other bands far and wide, integrating society infinitely. A widow would become a wife, extra or not, to an appropriate man, and so no one was unmarried or uncared for. We might pause to reflect on a large tribe inhabiting North America between the Atlantic and the Pacific.

Religion is a third kind of behavior that suggests itself strongly for the sapiens explosion. As with social forms, religious forms vary enormously. What is common to all is the strength of attitude, which is so great as to allow the supposition that it has a basic human component — that is to say, a biological one. It does great good for people, as individuals or as groups, and so it can be supposed to have evolved for the needs of modern humanity. As it probably existed in Neanderthals, this is not an idle assumption. If it has produced ugly and cruel history at times, this can be charged to human social nature, which one can argue calls for ingroup loyalty and outgroup suspicion and hatred. Again, our evolved social proclivities are probably adapted to the human groups of the late Pleistocene, before the rush to the large communities of today.

If religion expresses the highest philosophy and some of the greatest art of the last few millennia, this does not mean that it waited for the revelations that produced these things. The natural religious disposition of human beings is something that evolved, as modern people evolved. Creationist Christians may thunder against this, and against all evolution. But the very fervor with which they do this is, itself, a product of evolution.

We are fond of cosmic questions. Where did we come from? Why are we here? The answer is evolution. That may be a sufficient answer for some, though certainly not for most. But it is a necessary answer. There is no separating life — our own included — from evolution.

Glossary

Acheulean. The dominant stone-working tradition of the Lower Paleo-lithic (Europe, western Asia) and Earlier Stone Age (Africa) marked especially by bifacial hand axes, following the Oldowan.

Adaptation. Improved fitness of a species for a particular environment, acquired through natural selection.

Aegyptopithecus. A well-known fossil form from the Oligocene of the Fayum, small and with a tail, but possibly a forerunner of later hominoids.

Allen's rule. Warm-blooded animals living in colder parts of a habitat tend to have shorter limbs than their relatives in warmer parts, thus exposing less body surface to heat loss.

Amniotes. A term to group reptiles, birds, and mammals, all of which form a membrane, the amnion, to envelop the fetus, whether in an egg or the womb.

Ankarapithecus. A fossil ape from the Miocene of Turkey with cranial and dental characteristics apparently closer to hominids than those of known apes, past or present.

Anthropoidea. The "higher primates," comprising monkeys, apes, and hominids, living and fossil. Of higher organization and larger size, they appeared later in evolution (apparently in the Eocene). They have a continuous free upper lip, lacking the split lip and moist muzzle which prosimians (except tarsius) share with most other mammals.

Ardipithecus ramidus. The earliest species of *Australopithecus* currently known, with an age of almost 4.5 million years. Represented by some teeth and by jaw and skeleton fragments found in Ethiopia, it is clearly closer in form to a chimpanzee than are other australopithecine species and is thought to be close to the ancestral split between chimpanzees and humans. For this reason it is given a genus distinct from *Australopithecus* with the subfamily Australopithecinae.

Aurignacian.   The culture introducing the Upper Paleolithic to Europe, consisting of a recognizable set of stone-flaked blades plus numerous (bone) tools and objects of personal adornment. The point of origin is not clear but the culture is first known about 43,000 B.P. in eastern Europe, completing the occupation of western Europe by 35,000 B.P., replacing the Mousterian cultures made there by Neanderthals.

Australopithecinae.   The subfamily, within the family Hominidae, containing species assigned to *Australopithecus*, *Paranthropus*, and *Ardipithecus*.

Australopithecus.   The genus of early hominids (as known from over 4 million to 1 million B.P.) before the appearance of the genus *Homo*. The genus contains at least four species: *anamensis*, *afarensis*, *africanus* (the "gracile" forms), and *boisei*, the "robusts," which some place in a separate genus, *Paranthropus* (q.v.).

B.P.   Before the Present. 5,000 B.P. is the same as 3,000 B.C. in this book, but with ages of the order of about 20,000 years and more, reference to "B.C." becomes inappropriate.

Bicuspid.   The two teeth, better known as premolars, which lie between the canines and molars. Two-cusped in ourselves, the anterior lower premolars have a shearing edge in apes which bears against the upper canine.

Bilophodont.   "Two-crested teeth," describing the cross-crested pattern of the cheek teeth in Old World monkeys.

Brow ridges.   Constituting a bar of bone over the eyes in apes and early hominids, these are varied in development in later hominids, usually with internal sinuses. They are diminished in modern people to slight-to-moderate bony swellings (they do not correspond to the eyebrows), which are divided into a middle and lateral portion over each eye.

Cambrian.   The first epoch, beginning 590 million years ago, of the Paleozoic era, marked by the appearance of many new life forms with fossilizing hard parts. First vertebrates are found early in the Cambrian.

Canine fossa.   A sunken area in the cheekbone below the eye socket reflecting the retraction and vertical orientation of the modern facial skeleton. In the projecting midface of Neanderthals this fossa is not present.

Carbon-14 (radiocarbon) dating.   Based on the proportion of radioactive carbon (^{14}C) detectable in an organic sample (wood, bone, etc., especially burnt or charred). Carbon-14 is produced in the atmosphere by cosmic radiation but decays to nitrogen 14 at a regular rate so that the atmospheric ratio of carbon-14 is fairly constant. All living things incorporate carbon of both forms, inert and radioactive, but at death this process stops and the radiocarbon begins to decay with a half-life of 5,730 years (half decays in that time, half of the rest in another 5,730 years, and so on). This allows a good estimate of the elapsed time up to about 40,000 years, when measurement of remaining carbon-14 becomes difficult.

Catarrhine. "Downward-nosed." Applied to the Old World division (monkeys, apes, and hominids) of the Anthropoidea or higher primates, as distinguished from the Platyrrhines, or New World monkeys.

Cenozoic. "Recent life," or the Tertiary era ("Age of Mammals") beginning at 65 million B.P.

Cercopithecoidea. The superfamily which contains all the Old World (catarrhine) monkeys, distinguishing them from the remaining catarrhines, the superfamily Hominoidea, which contains the families Pongidae (apes) and Hominidae (hominids).

Châtelperronian. About 36,000 B.P., a short-lived western European stone culture of Upper Paleolithic type, probably used by Neanderthals under the impact of the Aurignacian culture.

Chromosomes. Paired sections of the DNA in the nucleus of a cell. The number varies in different species: in *Homo* the number of pairs is 23.

Coelacanths. A branch of the lobe-fins living in seawater and surviving longer than the freshwater lobe-fins. Largely extinct in the Mesozoic but one or two species still present.

Continental drift. The gradual shifting of continental plates in the earth's crust, which has caused such events as the separation of Africa and South America, the opening of the Rift Valley in Africa, etc.

Cro Magnon. A small cave shelter at Les Eyzies, in the Dordogne of France. Excavated in 1868, it produced the first skeletons of Upper Paleolithic Europeans, demonstrating their modern form. The name has become generalized to denote Europeans of the early Upper Paleolithic.

Dinosaur. Popularly used to mean the giant reptiles of the Mesozoic era ending 65 million years B.P., with tyrannosaurs, brontosaurs, and stegosaurs being familiar forms. Dinosaurs were actually only one division of the large "ruling reptiles," distinguished by features of the pelvis, and containing all the bipedal reptiles. Both dinosaurs and the other branches contained large and small forms; and ichthyosaurs (porpoise-like) and pterosaurs (flying reptiles) were not dinosaurs.

DNA (deoxyribonucleic acid). The carrier of the genes. It consists of extremely long paired sugar-phosphate chains (the "double helix") which are joined by pairs of four kinds of organic bases. The order of these last gives the codes by which the genes (segments of the DNA chains) control the formation of proteins.

Dryopithecus pattern. The pattern of cusps and fissures on the molar teeth common to all hominoids. Upper molars show four cusps, lower molars five: two on the tongue side, three on the cheek side, with a prominent Y-shaped groove embracing the middle cusp on that side; there is also a shallow transverse groove at the front of the tooth. The pattern clearly distinguishes teeth (human or ape) from the bilophodont molars of cercopithecoids (Old World monkeys).

Dryopithecus. A well-known genus of fossil apes first discovered in the mid-nineteenth century. Several species are recognized by different authorities.

Electron spin resonance (ESR). A radiometric dating method useful with some materials formed by precipitation, like clamshells, coral, or tooth enamel. Like thermoluminescence it depends on the release of trapped electrons accumulated from natural radiation but involves more complicated steps in this accumulation and methods of detection.

Eocene. "Dawn of recent." The second period of the Tertiary, from 53 to 36 million B.P. Prosimians abundant, especially in North America and Europe, with first signs of primitive higher primates, or anthropoids.

Epiphysis. In mammals, as the bony skeleton forms in cartilage ("gristle"), the joints of limb bones develop as separate bony pieces from the shaft (or diaphysis), thus allowing continued rapid growth in the cartilage between diaphysis and epiphyses before they unite at maturity.

Evolution. Species change over time, or "descent with modification."

Fayum primates. Fossil-rich sites in the Fayum depression of Egypt have yielded a number of Eocene and Oligocene fossil anthropoids, possibly containing ancestral forms of several later lines of higher primates.

Foramen magnum. The quarter-sized opening in the base of the skull through which the spinal cord passes. Joint surfaces for support of the skull on the spine are placed on either side.

Gene. A region of DNA on a chromosome, i.e., a sequence of base pairs, controlling the formation of a particular protein molecule. Genes accordingly act as pairs in this function, one on either chromosome. Location of genes to particular positions on human chromosomes is proceeding rapidly.

Gene flow.   Exchange of genes between populations through interbreeding at their borders, affecting gene pools in this way as opposed to through movements and replacements of population.

Gene pool.   The total of a population's genes, as to kind and proportion, determining the physical nature and the variation of that population. Individuals and family lines are samples drawn from the gene pool.

Genus (plural genera).   In formal classification, the first part of a binomial, e.g., *Homo*. A genus may contain a single species or several similar species; it has no definite limits of inclusiveness.

Gradualism.   The hypothesis that species change gradually, not episodically as in punctuated equilibria; sometimes called Darwinian evolution.

Hominidae.   A family of the superfamily Hominoidea comprising the genera *Homo*, *Ardipithecus*, and *Australopithecus* (the latter including *Paranthropus* of some authors). Hominids are characterized by a skeleton adapted for bipedal erect walking, which is reflected throughout the skeleton and skull, although the latter may be otherwise small-brained and ape-like except for small canine teeth. All hominids are extinct except *Homo sapiens*.

Homininae.   A subfamily (containing only *Homo*) of the family Hominidae as distinguished from the subfamily Australopithecinae of the same family.

Hominoidea.   The superfamily which contains the families Hylobatidae (gibbons), Pongidae (apes), and Hominidae (human beings and australopithecines), distinguishing all these from the superfamily Cercopithecoidea, or Old World monkeys.

Homo erectus.   The species recognized as including fossils dated from about 1 million to 500,000 B.P. (or later) from Africa, Java, and China, with brains varying around 1,000 cc and with robust skulls but skeletons generally modern in size and shape. Usually assigned regional subspecies, e.g., *H.e. leakeyi, H.e. erectus, H.e. pekinensis*.

Homo ergaster.   A species name used by some scholars to distinguish certain East African fossils (e.g., ER 3733, ER 3883, WT 17000) dated from 1.7 million to 1 million B.P. which are close to later *Homo erectus* but lack the robustness and some other features of that species.

Homo habilis.   The earliest recognized species of *Homo*, appearing at 2.4 million B.P. in East Africa and associated with the first recognizable stone tools. Distinguished from australopithecines by enlarged brain and reduced face; the skeleton, however, retained primitive traits not seen in later *Homo*.

Homo heidelbergensis.   A species proposed for "archaic" humans succeeding *Homo erectus* (e.g., the very similar specimens of Broken Hill in Africa and Petralona in Europe), giving rise to *Homo sapiens* and *Homo neanderthalensis*, respectively, in those continents. The name derives from the Mauer jaw found in 1908 near Heidelberg, as the first fossil of such a group.

Homo neanderthalensis.   The formal name for Neanderthals, recognizing them as a distinct species.

Homo rudolfensis.   An East African species contemporaneous with *Homo habilis*. Recognized as such by scholars who find the wider face and some other differences from *habilis* too great for the variation of a single species. Of interest because of the suggestion of rapid evolution occurring in *Homo* at this time.

Homo sapiens.   The formal species name for living mankind and for fossils recognized as essentially modern in form. It is distinguished from *Homo erectus* and *Homo neanderthalensis*, both of which survived after *H. sapiens* appeared, and from *Homo heidelbergensis*, a probable ancestor.

Humerus.   The bone of the upper arm.

Hylobatidae.   A family of the Hominoidea containing the gibbons.

K-T boundary.   At 65 million years B.P., the transition from the Cretaceous to the Tertiary was marked by some kind of catastrophic event and saw the extinction of the dinosaurs.

Lumbar spine.   The non-rib-bearing part between the ribcage and pelvis, usually consisting of five large vertebrae.

Mammal-like reptiles.   A branch of early reptiles remaining four-footed and developing various mammal traits of the skeleton. Very common at the Paleozoic-Mesozoic border but, except for mammal descendants, they became extinct as the dinosaurs developed.

Mandible.   The lower jaw, a single bone in mammals. Because of its density it survives well as a fossil, and because of its different shapes in hominids it is a good index of identity: in Neanderthals it typically lacks a chin and has a retro-molar space.

Marsupial.   These mammals do not form a placenta to nourish the fetus, but bring it forth in an immature state and nourish it in a pouch (marsupium). Except for a few intrusive species, all the mammals of Australia are marsupials; a few like the opossum are found in the Americas.

Masseter muscle. A short, quadrangular muscle between the cheek arch and the lower edge of the jaw, supplying power to the bite.

Mastoid process. A downward-pointing mound of bone behind the ear providing attachment for a muscle to the breastbone. With the vertical poise of the head in modern man, this muscle gets considerable play, and the mastoid process is typically more developed than in other hominids, including Neanderthals.

Mesozoic. "Middle life," or the "Age of Reptiles," running from 249 million to 65 million years B.P. The subdivisions are Triassic, Jurassic, and Cretaceous. It saw the dominance of the dinosaurs and other major reptile forms and the emergence of mammals.

Miocene. "Minority of recent." The fourth division of the Tertiary, from 18 to 5 million B.P. Monkeys and numerous fossil apes; appearance of hominids probably toward the end.

Mitochondria. Small elements within a body cell which use oxygen to convert foodstuffs into adenosine triphosphate which is used as energy by all parts of the body. Each mitochondrion has a small section of DNA (mitochondrial DNA, or mtDNA), far smaller than the DNA in the cell's nucleus.

Mousterian. A Middle Paleolithic assemblage of stone cultures of Africa, Europe, and western Asia comprising flake tools (scrapers and points), associated in Europe with Neanderthals but also with other hominids in Africa and the Near East.

Multiregionalism. The hypothesis of modern human origins which sees evolution of regional varieties ("races") as progressing separately from the stage represented by *Homo erectus*.

Multivariate analysis. A procedure for finding the best-defined differences among groups or individuals from their measurements or variables. Using all the material at once, thus preserving the total information, the methods transform this information into the best new axes separating such groups. The mathematics are not especially complex but because of the amount of computation involved the methods could not easily be used before the development of high-speed computers after World War II.

Mutation. Some change in the DNA, usually a change in one base pair or several, altering a gene's effect, minor or major.

Oldowan. An early Lower Paleolithic African culture of simply flaked pebble tools, appearing about 2.4 million B.P.

Oligocene. "Few recent." The third division of the Tertiary, from 36 to 23 million B.P. A number of early higher primates, known especially from the Egyptian Fayum; first South American monkeys.

Order. In classification of animals, a phylum is divided into classes, which are divided into orders. The class of mammals includes some fifteen orders, depending on the authority. We belong to the order of Primates which extends down to the lemurs.

Oreopithecus. Fossil hominoid from the Miocene of Italy. Structured somewhat like a small orang utan, it is notable for its aberrant teeth, making classification within or outside of the Hominoidea problematical.

Ouranopithecus. A Miocene hominoid from Greece, interesting as having some facial and dental traits closer to those of hominids than are the same traits in chimpanzees.

Paleocene. "Old recent." The first period of the Tertiary, from 65 to 53 million B.P.

Paleomagnetism. Polarity of the earth's magnetic field shifts over long intervals from north (as at present) to south and back again. Fine sediments and volcanic rocks preserve the polarity of the time when they formed, and so a scale of these events has been built up extending far into the past. The present "normal" phase began at 730,000 B.P. (or 780,000 by one recent measurement) and the previous "reversed" phase at about 1.7 million B.P.

Paleontology. The study of ancient life, based to a major extent on fossils.

Paleozoic. "Old life," or the "Age of Fishes," beginning at 590 million B.P., the first era, starting with the Cambrian period, marked by abundant fossils, including vertebrates.

Pan. The genus name for the chimpanzees. Some would include the closely related gorilla in the same genus; others recognize that *Pan*'s closest relative is actually *Homo.*

Paranthropus. The genus name for the "robust" australopithecines used by anthropologists who would separate these from the gracile species.

Pithecanthropus. The genus name selected by Dubois for his find of Java Man. It continued in use for the Far Eastern fossils for many years until sunk into *Homo erectus.*

Placental. The main group of mammals, those that form a placenta through which the fetus is nourished from the mother's blood stream.

Platyrrhine. "Flat-nosed." Applied to the New World monkeys to distinguish them from the Old World higher primates.

Pleistocene. "Almost recent." The period beginning 1.7 million B.P. and ending about 10,000 B.P. with the last glacial retreat; loosely called the "Ice Age." *Homo erectus* (or *Homo ergaster*) first appears at the beginning of the Pleistocene.

Plesiadapids. Numerous genera of earliest Tertiary primate-like animals in North America and Europe, but evidently not ancestral to primates generally.

Pliocene. "Majority of recent." The fifth division of the Tertiary, from 5 to 1.7 million years B.P. Monkeys expand; many apes extinct; hominids evolving in Africa.

Pongidae. A family of the superfamily Hominoidea containing the three large living apes.

Pongo. The genus name for the orang utan.

Potassium-argon dating. Radioactive potassium (^{40}K) decays to the inert gas argon (^{40}Ar) at a regular rate, and the two forms can be measured as trapped in rock. When a rock is heated to a very high temperature (as in volcanic lava) all argon gas is driven out, and new argon-40 begins to accumulate afresh. The time since a lava flow can thus be read in years, and since the half-life of potassium-40 is long, the scale is good for great distances into the past, but is useful only where volcanic rocks are found, as in East Africa.

Pre-adaptation. An accidental leg up in the process of adaptation through the previous existence of an organ or property which becomes converted to a new function.

Prosimii. The "lower primates," comprising lemurs, lorises, and tarsius. So termed because of the use of Simii, or simians, to denote the higher primates, properly known as the Anthropoidea. Prosimians are now found only in Africa and Asia.

Punctuated equilibria. The hypothesis that much species evolution takes place in episodes of rapid speciation with periods of stasis once new species are established. Opposite of gradualism.

Radius. One bone of lower arm, which does not partake in the hinge joint at the elbow; the circular upper end rests against the humerus and rotates against the ulna, so that at its lower end the hand can be turned through 180°.

Recessive gene. In Mendelian genetics a gene, in an unlike pair, which does not manifest its effect in the presence of its dominant partner.

Replacement. A term for the hypothesis of modern human origins which sees evolution to *Homo sapiens* occurring in Africa, followed by a relatively recent dispersal of populations to other regions and replacement by these of surviving earlier populations (e.g., *Homo erectus*).

Retro-molar space. In Neanderthal lower jaws, the protrusion of the face and thus of the tooth row leaves a space between the last molar and the ascending branch of the jaw when seen in profile.

Sacrum. Triangular block of fused vertebrae forming the back part of the pelvis, articulating with the lumbar spine and transmitting the weight of the upper part of the body to pelvis and legs.

Sickle cell trait. Due to a mutation, the sickling gene causes an abnormal form of hemoglobin in the red blood cells. This form is less efficient in the transportation of oxygen, and causes deformed shapes in the blood corpuscles, especially a sickle-like shape. These cells are, however, more resistant to invasion by malaria parasites; hence, an individual with one normal and one sickling gene, having a mixture of normal and sickling red blood cells, is at an advantage in populations exposed to endemic malaria. By natural selection, the proportion of the sickling gene rises to fairly high levels in such populations in tropical Africa. As the gene becomes frequent, the chances increase of getting a double dose of the gene, one from each parent, with no normal gene; this state results in sickle-cell anemia, usually fatal in early life, causing a fall in the sickling gene frequency, which cannot approach 100 percent. This situation is known as a "balanced polymorphism," in which opposing selections maintain a balance of the two genes. In the African-derived population of North America, with no selection pressure from malaria, the frequency of the sickling gene has been observed to be falling.

Sinodont. A term applied by Christy Turner to a complex of dental traits characteristic of mainland East Asia and American Indians. Such traits are more numerous and varied than those of the Sundadont complex (q.v.), so that Sinodonty seems like an intensification of Sundadonty.

Speciation. The establishment of one or more new species, with reproductive isolation, with or without survival of the parent species. This is a crucial event in the appearance of new species. Recognition of subspecies is a problem in *Homo*. It was also a problem for Darwin, who without genetics did not understand how a new species became fixed, in spite of titling his great work *On the Origin of Species*.

Species. In formal classification, the second part of a binomial, e.g., *sapiens* in *Homo sapiens*. In animal life it is the basic unit, definable as a set of populations that may interbreed successfully while maintaining its identity through its reproductive isolation from other species. It is also the basic unit in evolution.

Subspecies. A subunit of a species that can be distinguished physically and geographically from other subspecies, with which it can nevertheless interbreed easily. Geographical variation in modern humanity is not considered sufficiently well defined to justify subspecific distinctions, though there have been such attempts.

Sundadont. A term applied by Christy Turner to a complex of dental traits not found in Africa, Europe, or Australo-Melanesia, characteristic of populations peripheral to East Asia: Southeast Asia, offshore archipelagoes, Ainus, Pacific Islands.

Temporal muscle. A fan-shaped muscle attached to the upper point (coronoid process) of the jaw and spreading upwards on the side of the skull. Its area can be seen on the dry skull, indicating the power of the jaw in that individual or species.

Tertiary era. The "Age of Mammals" beginning 65 million years B.P., also known as the Cenozoic. Its divisions are Paleocene, Eocene, Oligocene, Miocene, Pliocene.

Tetrapod. Four-footed. A term to denote jointly, as the Tetrapoda, all the land vertebrates, from amphibians upward, distinguishing them from fishes.

Thermoluminescence (TL). A method of radiometric dating based on measuring the amount of electrons released when certain materials are experimentally heated. Burned flint (thus stone tools) heated in the past, thus driving off trapped electrons and allowing fresh accumulation of electrons from natural radiation, is an important material. Sand and other sediments are also useful, with heating by sunlight removing electrons at the time of deposition to give a clean slate in this way.

Ulna. One bone of the lower arm, forming a hinge joint at the elbow with the humerus. These two bones are involved in bending the arm.

Upper Paleolithic. In Europe, traceable from about 40,000 B.P., it is marked by varied tools worked on skillfully struck stone blades; also by bone tools and by art objects. It is associated with anatomically modern people and corresponds in time and general nature to the Later Stone Age of Africa.

Uranium-series dating. A radiometric method depending on the decay of radioactive isotopes of uranium and thorium to lead. Of variable usefulness, it can be applied to carbonates formed, for example, in mollusk shells and coral.

Vertebrates. The grand division of life, or phylum, that includes all back-boned animals: fishes, amphibians, reptiles, birds, and mammals. The formal name of this phylum is Chordata because it also contains a few varied forms that, like vertebrates, form a fibrous rod, the notochord, which precedes the spinal chord in vertebrate development, and that share a few features of arrangement with vertebrates. However, some chordates lack jaws, bones, or sense organs, etc., and above all lack the mobility which is a vertebrate's main character.

Zinjanthropus. The genus name first given to the robust "Dear Boy" of Olduvai Gorge. Now known as species *boisei* under either *Australopithecus* or *Paranthropus*.

Credits

Figures
(Illustrations are original unless otherwise noted.)

Fig. 1.   **Dubious Ancestors**, by A. Forestier, courtesy of *The Illustrated London News*.

Fig. 4.   **Skull Bones in a Lobe-fin Fish and an Early Amphibian** after William K. Gregory, *Our Face From Fish to Man*, Putnam, © 1929 by The Putnam Publishing Group.

Fig. 6.   **Skeletal Changes Leading to Mammals**, from Alfred S. Romer, "Major Steps in Vertebrate Evolution," *Science*, 12/29/67, vol. 67, © 1967 by AAAS.

Fig. 8.   **Reptile and Mammal Teeth**, after A.S. Romer, *Man and the Vertebrates*, University of Chicago Press, 1933, pp. 110, 118.

Fig. 12. **Wider, Please: A Baboon Shows His Teeth**, baboon after photo by Stanley Washburn in *The Primates*, Time-Life, 1965, p. 117; human mouth cavity after chart by A.J. Nystrom & Co., Chicago, Ill., reproduced in *Atlas of Human Anatomy*, Barnes & Noble, 1942, p. 51.

Fig. 13. **Body Proportions in the Great Apes and Man**, after A. H. Schultz, *The Life of Primates*, Weidenfeld and Nicolson, 1969, fig. 7.

Fig. 14. **A Young Orang Walking on His Balled-up Fists**, after F. Jenkins, *Primate Locomotion*, Academic Press, Orlando, Fla., 1974, p. 315.

Fig. 15. **Weight-bearing: Spine to Pelvis to Leg**, above drawn after E.A. Hooton, *Up From the Ape*, Macmillan, 1931, p. 109; below modified from Robinson, Freedman, and Sigmon in *Journal of Human Evolution*, 1972, p. 364, with permission of Academic Press, Inc., London.

Fig. 16. **Human and Ape Foot Skeletons**, after A.H. Schultz, *Folio Primatologia*, vol. I, no. 3-4, 1964, S. Karger AG, Basel.

Fig. 19. **Some Tertiary "Hominoids"**, after Michel Garcia in Yves Coppens, *Le Singe, L'Afrique et l'Homme*, Fayard, Paris.

Fig. 22. **The Laetoli Tracks**, photo of trail by John Reader, from Photo Researchers, Inc., New York; photo of footprints courtesy of Michael Day.

Fig. 23. **A Chimpanzee Walking Upright**, after photo by A. Kortlandt in *The Primates*, Time-Life, 1965, p. 71.

Fig. 24. **Brain Size, Jaw Muscles, and Skull Shape**, occipital views after drawings by Luba D. Gudz; jaw muscles from William Howells, *Mankind in the Making*, Doubleday, 1959.

Fig. 25. **The Afar Pelvis and Trunk**, courtesy of Kevin D. Hunt.

Fig. 26. **The Afar Skeleton in Two Postures**, courtesy of Henry M. McHenry.

Fig. 27. **Upper Jaws in Gorilla, *Australopithecus afarensis*, and *Homo***, gorilla and human after Howells, *MITM*; *A. afarensis* after Richard Klein, *The Human Career*, University of Chicago Press, 1989, p. 146.

Fig. 28. **The Skull and Palate of a Robust Australopithecine**, after illustrations in *Olduvai Gorge*, vol. II, Cambridge University Press, 1967, courtesy of Phillip V. Tobias.

Fig. 29. **Specimens Assigned to *Homo habilis***, after Klein, *THC*, p. 156.

Fig. 30. **A Family Tree for Hominids**, courtesy of Bernard Wood, reprinted with permission of *Nature*, February, 1992, © 1992, by Macmillan Magazines, Ltd.

Fig. 31. **The Nariokotome Skeleton**, after photo by Alan Walker in Richard Leakey and Roger Lewin, *Origins Reconsidered*, Doubleday, 1992; courtesy of Alan Walker.

Fig. 32. **Homo erectus Skull ER 3733**, after Klein, *THC*, p. 195.

Fig. 33. **Olduvai Hominid 9**, after photo by Tobias and Louis Leakey in Michael H. Day, *Guide to Fossil Man*, London, Cassell, 1965, p. 173.

Fig. 34. **The Best Java Man Skull**, after photo by author.

Fig. 35. **Java Man in Life**, after Gerasimov.

Fig. 37. **Far Eastern *Homo erectus***, all after Howells, *MITM*.

Fig. 38. **Peking Man in Life**, female after Jia, male after Gerasimov.

Fig. 39. **Progress in Stone Tools During the Paleolithic**, after Paul Mellars, "Major Issues in the Emergence of Modern Man," *Current Anthropology*, vol. 30, no. 3, June, 1989.

Fig. 40. **The Broken Hill Skull and Broken Hill Man in the Flesh**, skull after photo by author; head as restored by Gerasimov.

Fig. 41. **Post-erectus Europeans: Petralona and Arago**, Petralona after photo courtesy of Rupert I. Murrill; Broken Hill after Howells, *MITM*; Arago after photo in Day, *GTFM*, courtesy of Henry de Lumley.

Fig. 42. **The Dali Skull**, after photo by Milford H. Wolpoff appearing in April 1992 issue of *Scientific American*, p. 79, courtesy of Milford Wolpoff.

Fig. 43. **Two Borderline Skulls, Dated to About 130,000 B.P.**, Omo 1 after photo by Tobias in Day, *GTFM*, p. 244; Ngaloba (Laetoli Hominid 18), from photo by Mary Leakey in Day, *GTFM*, p. 189.

Fig. 45. **A Restoration of the Neanderthal Head and Torso**, from Howells, *MITM*, after Burian.

Fig. 46. **Neanderthal and Modern Physique**, from J.-J. Hublin in *Pour La Science*, no. 64, February, 1983, p. 63.

Fig. 47. **The Steinheim and Swanscombe Skulls**, after Howells, *MITM*.

Fig. 49. **The St. Césaire Skull**, after Mellars, *Current Anthropology*, vol. 30, p. 351.

Fig. 50. **Early Upper Paleolithic Cultures in Europe**, after Mellars as above.

Fig. 51. **Northwest African Remains**, Jebel Irhoud 1 & 2 after photos by author; Dar-es-Soltan after photo in A. Debenath, *L'Anthropologie*, vol. 90, p. 237, courtesy of Musée de l'Homme.

Fig. 53. **Modern and Neanderthal in the Near East**, Skhūl V after Howells, *MITM*; Qafza after photo by author; Zuttiyeh after J. J. Hublin in *Pour La Science*, no. 64, February, 1983, p. 68; Amud after photo from Israel Antiquities Authority appearing in article by Eric Delson in *McGraw-Hill Encyclopedia of Science and Technology*, 6th ed., vol. 7, N.Y., 1987.

Fig. 54. **Some Restorations by Gerasimov**, after Gerasimov.

Fig. 56. **Pre-modern Crania from Australia**, Kow Swamp skull after photo in A.G. Thorne and P.G. Macumber, "Discoveries of Late Pleistocene Man at Kow Swamp, Australia," *Nature*, vol. 238, August 11, 1972, © 1972 by Macmillan Magazines, Ltd.; Keilor skull after photo in D.J. Mulvaney, "The Prehistory of the Australian Aborigine," *Scientific American*, March, 1966 p. 02, © Scientific American, Inc. All rights reserved.

Fig. 57. **Portrayals from Major Racial Populations**, Mayan head after photo from Peabody Museum, Harvard University.

Fig. 58. **Upper Paleolithic Europeans**, Cro Magnon after photo in *Origines de l'Homme*, p.124, courtesy Musée de l'Homme; Předmostí and Mladeč after photos by author.

Fig. 59. **The "Tree" from Mitochondrial DNA for 189 Individuals**, from L. Vigilant, et al., "African Populations and the Evolution of Human Mitochondrial DNA," *Science Magazine*, September 27, 1991, vol. 253, fig. 3, p. 1505, © 1991 by AAAS.

Fig. 60. **The Border Cave Skull**, after photo by Tobias in Day, *GTFM*, p. 331.

Fig. 63. **Late Pleistocene Skulls from China**, all after photos by author.

Fig. 67. **The Lumper's View**, from schematic by Michael Day in Day, *GTFM*, fig. 133.

Fig. 68. **Still Another View**, by Mirachi, in *The Wall Street Journal*, November 10, 1986, permission, Cartoon Features Syndicate.

Photos

Cover photo of Cro Magnon skull from *Origines de l'Homme*, courtesy of the Laboratoire de Préhistoire of the Museé de l'Homme, Paris.

1. **Raymond Dart and Robert Broom**, photo kindness of Frances Wheelhouse from her book, *Raymond Arthur Dart: A Pictorial Profile*, Transpareon Press, Sydney.

2. **Louis and Mary Leakey**, photo by Des Bartlett, from Photo Researchers, Inc., New York.

3. **Eugène Dubois as a young army medical officer**, photo from *De Evolutie van de Mens*, Natur & Techniek, Maastricht, Netherlands, with permission of Mrs. A. Hooijer-Ruben.

4. **Franz Weidenreich and Ralph von Koenigswald**, Fritz Goro, *Life Magazine*, © 1946 Time-Warner, Inc.